Mosquitoes management

Environmental issues and health concerns

P.I.E. Peter Lang

Bruxelles · Bern · Berlin · New York · Oxford · Wien

Cécilia Claeys

Mosquito management

Environmental issues and health concerns

EcoPolis
Vol. 31

This book receives financial support from Fondation de France, Labex-Med and Aix-Marseille University- IRD-LPED.

Translation and English editing of this book by Jocelyne Serveau and Jessica Edwards.

This publication has been peer reviewed.

International Academic Publishers
Brussels, 2019
1 avenue Maurice, B-1050 Brussels, Belgium
brussels@peterlang.com ; www.peterlang.com

ISSN 1377-7238
ISBN 978-2-8076-0241-0
ePDF 978-2-8076-0242-7
ePub 978-2-8076-0243-4
Mobi 978-2-8076-0244-1
DOI 10.3726/b15130
D/2018/5678/107
Printed in Germany

Bibliographic information published by "Die Deutsche Nationalbibliothek".

"Die Deutsche National Bibliothek" lists this publication in the "Deutsche Nationalbibliografie"; detailed bibliographic data is available on the Internet at <http://dnb.de>.

Table des matières

Introduction

Cécilia CLAEYS

Associate Professor, Aix-Marseille University, LPED (Laboratory Population Environment and Development) UMR 151 AMU-IRD

According to the World Health Organization (http://www.who.int/mediacentre/factsheets (updated, October 2017)), mosquitoes are responsible for roughly 750,000 deaths worldwide each year (60% from malaria). Four hundred years BC, Hippocrates described the symptoms of malaria, and the anti-malarial properties of quinine have been known to the Western world since the late 16th century when it was brought back from Latin America by Jesuits who quickly began to commercialize the medicine. But it was not until 1870 that Cuban researcher Carlos Finlay first hypothesized that the yellow fever virus was transmitted by mosquitoes – a hypothesis that was confirmed a decade later. Then came the research of Scotsman Patrick Manson, of Ronald Ross, a Britain born in India (winner of the Nobel Prize in 1902), and of Frenchman Alphonse Laveran (Nobel Prize in 1907) describing how mosquitoes vector different diseases. On the heels of this medical research, a vaccine against yellow fever was developed in the 1930s. While not perfect, vaccination has allowed yellow fever epidemics to be controlled. There is however still no vaccination against malaria, chikungunya or Zika virus, and the dengue vaccination is in its infancy. Moreover, the synthetic antimalarial drugs that have replaced quinine since the 1940s now face parasite resistance. The limited success of prophylaxis in the past and still today means that vector control (VC) remains the main prevention strategy, at times alongside more or less effective curative treatments.

In countries in the South, vector-borne diseases are still a heavy burden on the population and vector control is a cornerstone of prevention strategies. And yet, the implementation of control measures is often confronted with the silent nature of most of these diseases. A large majority of people infected by mosquitoes that vector pathogens (viruses, parasites, bacteria, rickettsia, helminths) may not have any symptoms at all. In the least developed countries of the South, many sick people never consult a physician. And since the first symptoms of all these diseases are flu-like, differential diagnosis is difficult without complex bioanalysis. In countries in the North, on the other hand, health issues related to mosquitoes disappeared in the mid-20th century alongside overall improvements in living conditions and public health, and mosquito management has since shifted towards ensuring the comfort of populations and tourism-related concerns. Much emphasis has been placed on the widespread development of insecticides, with funding from both public authorities and private actors.

Yet the combined effects of globalization, urbanization and climate change have begun to undermine the North-South divide in terms of the threat posed by mosquitoes (nuisance vs. human health) and the respective responses they receive (massive eradication vs. targeted VC). More frequent human travel and the increased circulation of goods have encouraged the transportation of mosquitoes and pathogens from one continent to another, while climate change and urbanization have facilitated their implantation in new territories (Hawley, 1988; Paupy *et al.*, 2009). Such global changes are not only transforming distribution ranges, they are also changing the genetic makeup of vectors (greater vector competence) and pathogens (greater virulence). This is the case, for example, with genotypic selection in the North-American West-Nile virus which has become more virulent for humans. In Africa, the M molecular form of *A.* Gambiae (a malaria vector) has adapted to urban pollution, thus making it more effective at spreading malaria in urban areas. Finally, we have also witnessed the selection and emergence of a new genotypal form of chikungunya better able to replicate when associated with *Ae. Albopictus.*

Most research into mosquitoes – particularly in the social sciences and humanities – has tended to embrace the North-South geographic opposition in its representations of vector-based risk. Until the early 2000s, there was a clear distinction between environmental- and health-related issues. Given that mosquitoes were no longer a health threat in the North, research tended to focus mainly on the environmental controversies surrounding comfort-based mosquito control. Conversely, in the South, research in the social sciences into the different diseases transmitted to humans by mosquitoes was mainly conducted in the context of the social sciences of health and the fieldwork sites selected tended to be large, malaria-endemic areas. And yet it has now become necessary to challenge this dichotomy between environmental concerns in the North and health concerns in the South – which is exactly what this volume aims to do.

Given that both health and the environment sit at the crossroads of nature, culture and technological and organizational evolution (Mougenot, 1998), it is essential to compare them. Already in his time, Hippocrates wrote in *On Airs, Waters and Places*, "Whoever wishes to investigate medicine properly, should [...] consider the seasons of the year, [...] then the winds, [...] the qualities of the waters, [...] the ground, [...] and the mode in which the inhabitants live" (Cicolella, 2010, quoting Hippocrates, *The Genuine Works of Hippocrates.* Edited by Charles Darwin Adams. New York Dover, 1868, p. 20). Beginning in the 19th century, modern medicine and the development of public statistics made it possible to measure the connection between human health and the quality of an environment (Fassin, 2009). The environment as a notion did not yet have its contemporary meaning, however, and was associated with "that which surrounds us", without distinction between anthropic and natural milieus and the social sphere. And yet historian Jean-Baptiste Fressoz (2013) has noted that during the 19th century, scholarly discourse tinged with hygienism for example tried to "deeply reshape medical etiology" in favour of a moralizing register. Deemed immoral, the behaviour of the sick was incriminated – be it tuberculosis among the urban European proletariat or yellow fever in slaves and their offspring in the colonies. This was a convenient way to blame the victims and elude the underlying causes which, ultimately, were the capitalist and slave-driven exploitation of poor populations concentrated in insalubrious housing in the zones most exposed to pollution and disease.

The emergence – or rather the re-emergence – of the health/environment debate is relatively recent (Carricaburu, 2005; King and Crews, 2013). In 1962, naturalist Rachel Carson condemned the effects of DDT (Dichlorodiphenyltrichloroethane) on ecosystems and human health in her book *The Silent Spring* which went on to symbolize the rise of contemporary Western ecological awareness. Regarding the effects of DDT in particular, the mediatisation of Carson's book helped expose a latent controversy (Gunter and Harris, 1998). The insecticidal properties of DDT were discovered by Swiss chemist Paul Hermann Müller in

1938 (awarded the Nobel Prize in 1948) and DDT was produced on an industrial scale to enable its widespread use starting in 1942. The American army used it abundantly for mosquito control in the Mediterranean and tropical regions during its post-war "liberation" campaigns. From 1950 to 1969, the World Health Organization recommended using DDT in its malaria prevention campaigns. DDT was also massively used in the agricultural sector and as a domestic insecticide. In 1972, the United States Environmental Protection Agency banned the use of DDT on American soil, but not its foreign exportation. In 2004, the Stockholm Convention added the substance to its list of persistent organic pollutants. DDT nevertheless remains in use in many countries in the South, fuelling recurring socio-technical controversies that pit human health against ecological concerns (Bouwman *et al.*, 2011). In the North, DDT has gradually been replaced with more selective and less residual molecules, but these are nevertheless potent products that are not neutral on ecosystems. Indeed, vector control on all continents involves a toxic intrusion into the ecosystem and, despite this, the insecticides used are not efficient at sustainably eradicating mosquitoes. It has recently been shown that mosquitoes are extremely good at fabricating biodegradation enzymes capable of resisting control treatments. Moreover, even when organophosphorus (i.e., less polluting) insecticides are used, mosquito control campaigns may: i) reduce the diversity of mosquitoes by eliminating some of the natural competitors of the mosquitoes that vector human diseases, and ii) destroy insects that are useful to mankind, such as pollinating bees. In both the North and South, public authorities and populations are now confronted with health problems and nuisances against which vector control has rarely been as efficient as hoped – and whose risks for the ecosystem and human health are increasingly obvious.

The difficult struggle against vector mosquitoes and its unwitting effects are a potent reminder of the interdependence between human health and the health of ecosystems. "Oversight" of this detail goes hand in hand with historical blindness to what are now called environmental inequalities – i.e., the compounding of social vulnerabilities experienced by certain populations and their exposure to degraded or at-risk environments. It was not until the emergence of a theoretical school and activist movement surrounding environmental justice that this issue was explicitly placed on the Western political agenda (Taylor, 2000; Gislason, 2013). The notion of environmental justice broadens the scope of social justice. To begin, it underscores that the search for social justice must include a decrease in social inequalities in terms of health *AND* exposure to environmental risks. Secondly, it argues that social justice and environmental protection can be complementary. The issue of vector mosquitoes fully engages with this type of thinking. It has repeatedly been shown that the most socially vulnerable populations are most affected by exposure to vector mosquitoes. Moreover, the difficult search for alternatives to the leading phytosanitary treatments used in vector control today puts to the test the potential for complementarity between the protection of human health and that of ecosystems (Mieulet and Claeys, 2016; Claeys and Mieulet, 2017).

This book takes an international approach to the interactions, interdependences and tensions between health, environmental and social issues surrounding mosquito control. Scientists and social scientists specialized in environmental issues examine (re-)emerging health issues and engage with specialists on health-related topics who are also interested in exploring growing environmental concerns.

In the first chapter, medical entomologist Florence Fouque provides an overview of her scientific career from the perspective of interactions between health, the environment and society. A specialist on vector mosquitoes, she underscores the extent to which our understanding of vector-borne diseases (VBDs) needs to consider the social and environmental factors at play. The chapter uses concrete examples to measure the success of such approaches and assesses the challenges ahead. The author places particular attention on ethical considerations in terms of scientific practice and operational actions in fighting VBDs.

In the second chapter, Mariam Otmani del Barrio examines the processes behind the compounding of health, environmental and social vulnerabilities regarding the exposure of populations to VBDs. She offers an overview of the global situation and points out how a complex intertwining of socio-economic, cultural and demographic vulnerabilities aggravates health risks. Turning the focus more specifically to South America, Otmani del Barrio develops a theory regarding the multiple vulnerability of women during the recent Zika epidemic.

The third chapter addresses the re-emergence of dengue in the United States (US). After a presentation of the general context in the US regarding mosquito-borne disease and control, Melinda Butterworth compares three regions: southern Florida, southern Texas and southern Arizona. This spatialized comparison makes it possible to identify the influence of socio-ecological factors that worsen or reduce the risk of vector-borne disease. The chapter really emphasizes the dynamic nature of socio-environmental processes related to VBDs which evolve over time and across space. This leads the author to underscore the importance of prevention policies that are reactive and able to adapt to ongoing socio-environmental changes.

The fourth chapter addresses the impact of climate change on the malaria transmission cycle. The analysis is focused on Burkina Faso. Using quantitative studies and modelling, Eric Diboulo shows how the most socially vulnerable populations tend to be the first affected by higher malaria transmission rates caused by climate change. He argues that it is important for malaria control policies to consider different climate change scenarios and emphasizes the particular vulnerability of sub-Saharan populations.

The fifth chapter looks at Italy, the first European country to be affected by a dengue epidemic in the early 2000s. Paolo Giardullo analyses Italian national media coverage of this emerging issue in environmental health. He notes the relative invisibility of the issue at the national level during the decade following the first epidemic in the Emilia Romagna region. Paradoxically, it was not until 2016 and the Zika crisis in Brazil in the lead-up to the Olympic Games that the Italian media clearly embraced the issue.

The sixth chapter presents a socio-history of mosquito control policies in mainland France and the French Antilles. Cécilia Claeys provides a postcolonial interpretation of the shift from vector control to comfort-based mosquito control. The chapter examines the decolonialization process and the greening of mosquito control policies on both sides of the Atlantic, and highlights the challenge in conceiving a socio-environmental and health-based community of destiny between the mainly white population of mainland France and the multi-ethnic overseas population comprised in part of the descendants of slaves.

The seventh chapter analyses the socio-environmental controversy surrounding the chikungunya epidemic that affected Reunion Island in 2005-2006. The chapter provides insight into the array of postcolonial situations that exist in the overseas territories of France. Author Marie Thiann-Bo Morel shows how a direct transfer of health-crisis-management strategies from the French Antilles became the focus of local protest which refused to sacrifice the environment in the name of a sanitary emergency.

The last chapter launches a call for more interdisciplinary research into the topic. Authors Cécilia Claeys, Valérie Bertaudière-Montes, Christine Robles, Magali Deschamps-Cottin and Julie Cardi build on the findings of a research programme involving sociology, plant ecology and medical entomology, and add in an architectural perspective. Based on a comparison between mainland France and the French Antilles, this interdisciplinary study examines the proliferation of mosquitoes that vector dengue, chikungunya and Zika in individual homes with gardens. Their results lead to the formulation of recommendations for housing that is both sustainable and anti-vectorial, and which need to be better included in vector control strategies to encourage crossover between the protection of the environment and human health.

References

Bouwman H., van den Berg H. and Kylin H. (2011). "DDT and malaria prevention: addressing the paradox". *Environ Health Perspect*, 119(6), 744-747.

Carricaburu D. (2005). « Santé/Environnement : des savoirs aux pratiques ». In Agence nationale de la recherche, *Santé-environnement et santé-travail : nouvelles perspectives de recherches*. Document d'orientation scientifique du ministère de la Recherche, Paris, 31 mars et 1er avril, 1-14.

Cicolella A. (2010). « Santé et environnement : la 2e révolution de Santé Publique », *Santé Publique*, 22(3), 343-351.

Fassin D. (dir.) (2009). *Inégalités et santé. Problèmes politiques et sociaux*, Paris, La Documentation française.

Fressoz J.-B. (2013). « Biopouvoir et désinhibitions modernes : la fabrication du consentement technologique au tournant des XVIIIe et XIXe siècles », *Revue d'histoire moderne et contemporaine*, 60(4-4bis), 122-138.

Gislason M.K. (ed.) (2013). *Ecological Health: Society, Ecology and Health (Advances in Medical Sociology, Volume 15)* Bingley (UK), Emerald Group Publishing Limited.

Gunter V.J. and Harris C.K. (1998). "Noisy Winter: The DDT Controversy in the Years before *Silent Spring*", *Rural Sociology*, 63(2), 179-198.

Hawley W.A. (1988), "The biology of Aedes albopictus. Review article", *Journal of the American Mosquito Control Association*, 1(Supplement), 2-39.

King B. and Crews K.A. (eds.) (2013). *Ecologies and politics of health*, London, Routledge.

Mougenot C. (1998). « Santé et environnement : analogie dans les approches sociologiques », *Biotechnol. Agron. Soc. Environ.*, 2(2), 96-105.

Paupy C., Delatte H., Bagny L., Corbel V. and Fontenille D. (2009). "Aedes albopictus, an arbovirus vector: from the darkness to the light", *Microbes and Infection*, 11(14-15), 1177-1185.

Taylor D.E. (2000). "The Rise of Environmental Justice Paradigm. Injustice framing and the Social Construction of Environmental Discourses", *American Behavioral Scientist*, 43(4), 508-580.

Vector-Borne Diseases and human societies: some examples of relationships, evolutions and challenges

Florence Fouque

Vectors, Environment and Society Unit, Special Programme for Research and Training in Tropical Diseases (TDR), World Health Organization (WHO), Geneva, Switzerland

Vector-Borne Diseases (VBDs) are clinical symptoms caused by pathogens transmitted by a living organism called a vector, to specific hosts such as humans and animals. These pathogens can be viruses, bacteria or parasites, including some multicellular organisms such as worms, and they are transmitted mostly by insect vectors. What is quite unique in the transmission mode of most VBDs is the need for blood-feeding behavior in the vector, even if some such diseases do have different transmission modes such as through the skin or by ingestion. Vectors may also include other arthropods or even snails and can also be affected by hosting pathogens (Wilson *et al.*, 2017). Systems that include a vector, pathogen, host and their environment, are called ***vectorial systems***. Pathogens can cause clinical symptoms for the host, but they can also circulate in a system without any clinical outcomes; VBDs thus have a very large range of impact on hosts and host populations, spanning from no disease and no impact on public health to deadly epidemics. Another peculiar feature is that a large range of potential impacts can be found within a single type of system, like for example in the system of dengue viruses, which includes only mosquito vectors, human hosts and dengue viruses. Dengue circulation into a population can range from completely silent and endemic to highly epidemic with huge numbers of severe cases. The factors behind the circulation of such viruses as well as their impact on human health are not all well known and thus impair efficient control of this VBD (Andraud *et al.*, 2012). Conversely, the cultural approach to VBDs is a fundamental element to be taken into consideration for the satisfactory mitigation of these diseases. Human societies have always had to contend with the challenges of VBDs, but disease transmission is also affected by the changes and structure of human societies, and the history of humanity has been strongly affected by such diseases (Dobson and Carper, 1996). As far back as 400 B.C. Thucydides reported a typhus-like epidemic in Athens that killed a quarter of the city's population and changed the face of the Peloponnesian war (*Figure 1*). The plague pandemic in Europe in the middle-ages changed the history and economy of the British Isles, and the Yellow Fever epidemic among the French troops sent by Napoleon to suppress the Haitian insurrection resulted in the independence of Haiti in 1804, the first example of independence following the colonial period. The mode of transmission of VBDs was only discovered and proven quite recently – the first evidence was found during the second half of the 19th century (Service, 1978). The proof came mostly from scientific and technical discoveries; the importance of social factors has indeed only come to the forefront more recently. VBDs are strongly tied to human behavior and when technical solutions are implemented without consideration for the social context, it becomes very difficult to obtain sustainable prevention and control against these diseases. In the following sections, I would like to show how social context affects the transmission of VBDs through different examples from situations in which I have been involved over the past 30 years.

Figure 1. *Illustration of the Athens' plague by Nicolas Poussin (1594-1665).*

When economic development helps control vector-borne diseases

The first example of the relationship between human societies and VBDs comes from southern Switzerland and the Canton of Ticino which was an endemic malaria area until the end of the 19th century due to the vast wetlands of the Magadino plain where the Ticino River enters Lake Maggiore. The climate of the Magadino plain is very hot and humid in the summer and subject to mild winters. The plain was a malaria endemic area until the wetlands were dried in the 1880s and the Ticino River was regulated to avoid widespread flooding. Such modifications resulted in the economic development of the area, where large agricultural fields could be exploited, and in the eradication of malaria alongside the disappearance of *Anopheles* mosquito breeding sites. In the first years of this development, the interaction between human activities and natural sites went smoothly, with only moderate impact on local biodiversity and an approach based on a kind of co-existence. This type of human/natural area relationship is common in "primitive societies", for which nature is part of the necessary environment. After the 1960s, economic development accelerated with the intensification of agriculture, the arrival of mass tourism and other industrial activities, followed by a loss in biodiversity resulting in an ecological movement to protect the remaining natural sites. Consequently, part of the Magadino plain was transformed into a natural reserve and protected area for migratory birds – and this resulted in the wetlands coming back, along with huge numbers of mosquitoes (Fouque *et al.*, 1998). In this context, a heavy nuisance developed which resulted in a decline in tourism revenues and an economic burden on the Canton of Ticino. The nuisance was unacceptable not only for the local population, but also because of the economic loss. When local authorities decided to address the problem, no information on the mosquito populations and species was available, however, and the causes of this nuisance outbreak were unknown. This situation is a good example of how a lack of basic knowledge about mosquitoes can lead to an uncontrolled situation, thus underscoring the need for continuous surveillance and collection of basic data to prevent and control outbreaks. The first step to control this nuisance was to identify the biting species, including the biting behavior and the species bionomics.

Insert 1: Ethical issues. To collect biting mosquitoes, the only technical solution is human landing collections, which raises ethical questions about this type of experimental collection. The definition of what is ethical and/or acceptable is not straightforward (Achee *et al.*, 2015). If mosquito collections are decided and undertaken by the same person are they more acceptable than when they are completed by paid staff supervised by someone else? If the collections are for a nuisance are they more ethical than when they are done for species transmitting diseases and, if collectors live in an endemic country, exposed to bites all the time, is it ethical to pay them for collecting vectors? There is unlikely a simple answer to such questions which are tied to social conditions, knowledge of the risks, acceptance of the consequences, the urgency of the situation and the availability of other methods. For all these reasons, the informed consent of participants is needed, and Ethics Committees are now in place everywhere. The discussion with and recommendations given by such committees are the best approach currently available, until new tools to collect mosquitoes are developed.

To resolve the nuisance in the Magadino plain, the biology and ecology of the biting species were studied. Modeling was further used to better understand how the different factors influenced fluctuations in the populations. This work resulted in hypotheses to explain why and how the nuisance developed and what solutions were available. Among the potential solutions some proposed environmental changes in the reserve area (Fouque *et al.*, 1998) while others recommended larvicidal control using the organic biocide *Bacillus thuringiensis var. israelensis* (*Figure 2*). Local authorities ultimately opted for control since the reserve area was protected and ecological changes would have impacted the natural landscape of the reserve. This situation provided a good opportunity to examine the relationships between mosquitoes and humans in the context of the protection of natural environments. Contradictory positions and conflicting interests are common and, in this case, the conflict was between economic interests and biodiversity. Human development and activities destroy vast natural areas worldwide, often forgetting that such areas are vital to human survival, as ongoing climate change shows. However, it is not easy to strike the necessary balance between human activities and protection of the natural environment. We lack the tools to determine when and where to set limits. In the Magadino plain, the mosquito issue has been resolved for over 20 years, but a new threat now exists with the arrival of the invasive *Aedes albopictus* mosquito species, a potential vector for many arboviral diseases. The species has invaded all urban and peri-urban environments and its population is still growing (Flacio *et al.*,

2015). No solution to control this new species has yet been found and surveillance and prevention are once again urgently needed to avoid future disease epidemics. With *Ae. albopictus*, we face new challenges tied to the domestication of mosquito vectors. This is well illustrated with the history of *Aedes aegypti* discussed in the next section.

Figure 2. *Evolution of the natural environment in the protected "Bolle di Magadino" area, (A) where increased forested and shaded areas created favorable breeding grounds for Aedes vexans and a land management proposal (B) to eliminate some of the breeding sites and to open a channel to give access to other breeding sites for predatory fish in order to reduce mosquito populations and the subsequent nuisance (extracted from Fouque et al., 1998).*

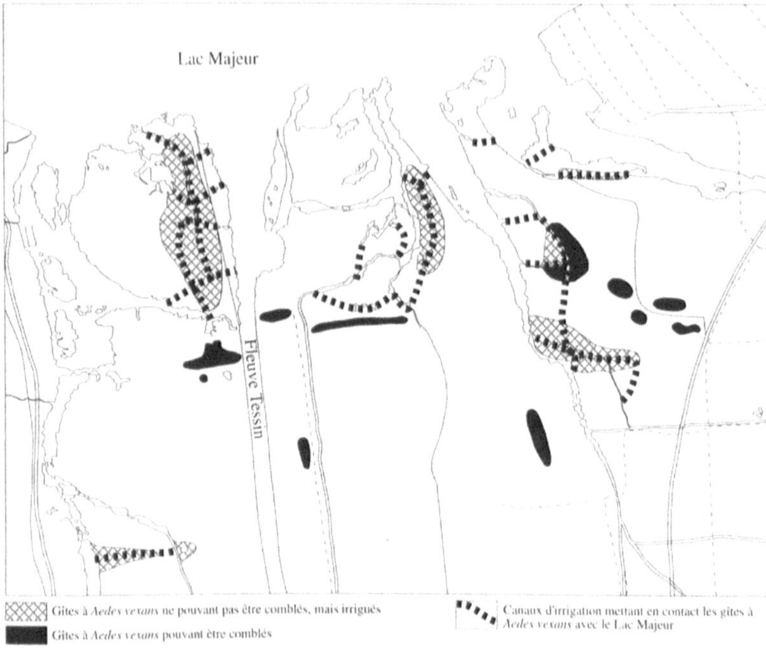

The domestication of Aedes aegypti and urban/peri-urban arbovirus transmission

My work conducting medical entomology investigations took me to French Guiana to study dengue vectors and the *Ae. aegypti* mosquito, following the largest dengue outbreak in recent history in this French department in South-America in 1992. The outbreak caused thousands of cases, including hemorrhagic cases and roughly ten deaths in a population of about 200,000 inhabitants. It was the first episode of dengue hemorrhagic fevers reported in South-America in the 1990s. Twenty-five years later, in 2017, dengue outbreaks are still occurring regularly in this territory and, while public health authorities now know how to clinically manage such cases, we still do not have efficient and sustainable solutions against the mosquito vectors. While vector control is successful in some other situations, for dengue transmission, we are still missing the target. The history of the relationship between the *Ae. aegypti* mosquito species and human society is a textbook example since the mosquito moved a hundred years ago from its original sylvatic environment in Africa to a domestic environment in and around inhabited areas. Although the mosquito has been extensively studied due to its high competency for transmitting some of the deadliest arboviral diseases such as yellow fever and dengue, we still lack efficient tools to control its population and disease transmission. This is paradoxical since the species lives within human societies. The shift of *Ae. aegypti* from a natural environment to a domestic environment has resulted in a complete modification of its ecology and biting preferences (Powell and Tabachnick, 2013), as well as many other features that are less visible such as its gonotrophic cycle because the species can take blood meals all the time, including when it is maturing eggs. In French Guiana like most South-American countries, *Ae. aegypti* was eradicated after an intensive campaign using DDT insecticide in the late 1940s. This eradication was to control yellow fever outbreaks at a time when the vaccine was not available for all. For the first time in recent history, human societies implemented an efficient tool to control mosquitoes with insecticide products and the success of mosquito control with such products indeed produced the best results. However, insects are very resilient organisms and some remaining pockets of *Ae. aegypti* populations became resistant to the insecticides as quickly as 10 years later, in the 1950s, and started to recolonize the continent, which was completely reinvaded within about 20 years, by the late 1960s. The recent building up of *Ae. aegypti* populations, which were already known as a very efficient vector for several viruses should have alerted the national health authorities to the risk of an arboviral disease outbreak. And yet nothing happened and, when the first modern dengue epidemic hit Cuba in 1981, it was considered a surprising event. It then took almost ten years before the first dengue outbreak in French Guiana, in 1992, but again the health system was not prepared.

Following these dengue outbreaks, vector control measures against *Ae. aegypti* were re-organized and, in many countries, help came from malaria control programs. Nevertheless, the fight against *Ae. aegypti* mosquitoes was developed in a difficult political context, with lack of trained staff and using outdated tools and products, without any long-term perspectives. The case of entomological surveillance tools offers a good example of this gap. The chosen tools were simply borrowed from yellow fever recommendations and were rarely tested against dengue transmission. Among the most used, the Breteau Index and the House Index based on larval surveillance, do not have any clear relationship with dengue transmission (Bowman *et al.*, 2014). Vector control is based on the spraying of insecticide products resulting in an increased mosquito resistance, without really demonstrating any efficacy against dengue transmission. Finally, emphasis is placed on community participation, based on the fact that *Ae. aegypti* breeds in domestic containers, and so the idea is that everyone in a community can manage the potential breeding sites in his or her own home. Community participation can work for a short period, but it is almost impossible for non-trained individuals to sustain checking all potential breeding sites, particularly during the rainy season in the tropics. Further, the living conditions of poor populations require huge efforts from them to keep their homes mosquito-free. And what about public spaces in which public authorities are also unable to control the mosquitoes? In this context, the dengue situation has continued to worsen across all Latin American countries and worldwide, leading to the current situation in which dengue is a threat to half the worldwide population, with more impact on the poorest populations most susceptible to VBDs (*Figure 3*).

Moreover, the tropical world (although not only) is now facing the emergence of other *Aedes*-borne arboviruses which pose a threat to public health globally such as the chikungunya and Zika viruses. Both diseases have severely affected millions of people in the American region, since it was a new territory for both viruses, with the arrival of the chikungunya virus in the Caribbean in 2013 and the arrival of Zika virus in Brazil in 2014 (Patterson *et al.*, 2016). With Zika virus, new clinical outcomes appeared with the impact of the virus on the nervous cells, and notably the unexpected increase in Guillain-Barré cases and microcephaly in newborn babies. Both viruses were first isolated in Africa, where they were considered mild because they did not result in human epidemic and/or severe cases. However, their impact on human health has dramatically changed in recent years, maybe because they are transmitted by different vectors, in different environments and have affected unprepared populations. The emergence of these "new" diseases is the direct consequence of the domestication of *Ae. aegypti*, and we are far from understanding all the potential implications of this domestication. Further, another mosquito species – *Ae. albopictus* – is currently following the same path of domestication.

To control *Ae. Aegypti*-borne diseases, we need to more thoroughly understand how the mosquito populations are influencing disease transmission. We know that several factors are involved including densities, longevity, biting preferences, infectious rates, and many others

(Goindin *et al.*, 2015), but we lack indicators to measure the force of the transmission. We already know that it is the oldest females that transmit the viruses, so why don't we use this mosquito stage to develop better indicators?

Finally, if we are looking at all the available methods to reduce the *Ae. aegypti* mosquito population to a level that is incompatible with dengue transmission, we can see that huge efforts are nevertheless yielding very poor results. The eradication era showed that an almost military-like approach to insecticide spraying was the most successful. But the results were not sustained and it is now impossible to apply the same method. New technologies are arriving such as the genetic modification of mosquitoes, but we are still awaiting proof of the impact of these methods on disease transmission. Further, the cost of such methods may not be affordable for the majority of countries burdened by dengue. How can we stop contact between the mosquito vector and the human host? That is what must be investigated because, as the following section shows, it is when such contact is stopped that disease prevalence decreases.

Figure 3. *Mean numbers of DALYs (Disability Adjusted Life Years) due to VBDs, with maximum yearly numbers between 2000 and 2015, plotted against the mean GDP (Gross Domestic Product) per capita (mean annual values for 2010-2016) in countries in the WHO regions. (Data on DALYs are extracted from http://www.who.int/healthinfo/global_burden_disease/estimates/en/index2.html, and data for GDP are extracted from http://data.worldbank.org/indicator/NY.GDP.PCAP.CD).*

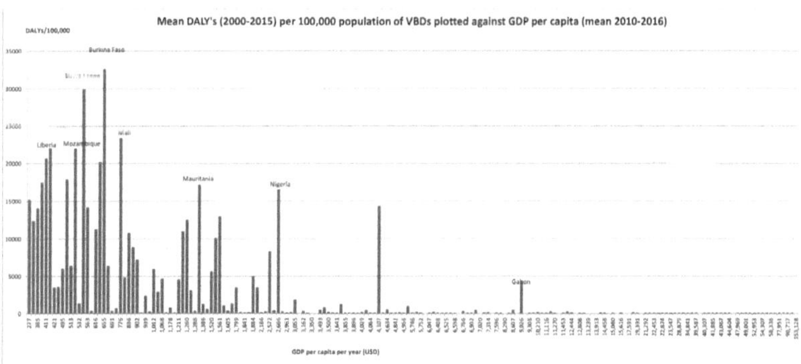

Malaria transmission and recent success

The number of malaria cases worldwide in 2015 was estimated at 212 million and, although the incidence of this disease has decreased by 41% since 2000 (WHO, 2016), it still represents the highest burden of VBDs in terms of clinical cases and deaths. While malaria is certainly not a recent disease – it was described by Hippocrates about 500 years B.C. –, its transmission cycles were unknown until Laveran in 1880, and the disease was associated with swamps until Ronald Ross in 1897 who was the first to prove the role of mosquitoes in malaria transmission (Neghina *et al.*, 2010). Malaria was common in the northern hemisphere and as far north as Sweden until the last century but it disappeared with the development and improvement of health care and facilities. While the discovery and production of chloroquine resulted in cutting the malaria cycle, the complete eradication of malaria in northern countries before the era of insecticide-based vector control is not fully understood (Hulden *et al.*, 2005). The malaria cycle can be interrupted more easily than dengue because there is a window between the outcome of clinical signs (fever) and the appearance of the gametocytes in the blood of the host. The gametocytes are the only forms of the parasite that can pass through the barrier of the midgut in the mosquito's body of a competent species and then allow the parasite to complete the cycle in the mosquito. This cycle is called the Extrinsic Incubation Period (EIP). Consequently, when a patient is treated at the onset of the fever, the parasites are killed, the gametocytes cannot develop and the mosquito cannot get infected or become infectious.

The malaria cycle was also cut in many tropical countries because of early diagnosis and treatment, coupled with vector control such as in French Guiana (Floch, 1954). Unlike the cycle of the dengue viruses, when the *Anopheles* vector species became resistant to insecticide, health facilities where still in place for diagnosis and treatment, and malaria did not come back. The first successes against malaria led to the global eradication campaign launched in the 1950s (Mendis *et al.*, 2009). However, in many poor countries and for the most neglected populations, this campaign was a big failure, particularly because health facilities were not in place to sustain malaria control and the disease came back with stronger epidemics. Further, around the 1960s when many tropical countries became independent, the withdrawal of the colonialist health infrastructure left countries in a very difficult situation, and VBDs including malaria came back with increased intensity. In the following years, the struggle against malaria was thought to be lost until a simple protective tool proved some effectiveness in stopping the contact between vectors and human hosts. This simple tool was an impregnated bed net or Long-Lasting Insecticide Net (LLIN) (Lengeler, 2000). For roughly the past 15 years, huge efforts have been made worldwide to deploy LLINs, coupled with some Indoor Residual Spraying (IDRS) – and the results are outstanding, with a decrease in malaria prevalence of 40% and malaria elimination already achieved in many countries (Cibulskis *et al.*, 2016). This global effort cost an estimated USD 4.35 billion for Africa alone between 2012 and 2014 (*Table 1*). The challenge is now to achieve this elimination in more countries and to address what is known as residual malaria which is due to changes in mosquitoes (other vector species, change of biting times and place) and changes in human behaviors. Social factors are now appearing more and more important in controlling malaria. Further, malaria is linked to poverty and political choices, and may persist only in countries that do not have the means to deploy the necessary health services. Even in the tropics, all High Income Countries (HIC) have eliminated malaria, the example of the Caribbean countries is very informative.

Malaria control is an example of success against a very old VBD. We know what can be done to control this disease and how the methods must be implemented. The persistence of the disease is now tied to social factors, such as poverty, civil unrest and the abandoning of neglected populations and ethnic groups by local public health authorities. There are numerous examples of small groups of populations that do not receive adequate health services and the importance given to these groups by the UN Sustainable Development Goals shows how the improvement of their health is necessary for global health.

Table 1. Estimated cost of malaria control in African countries compared to the revenues of the countries (data on costs were extracted from Annex 3; Funding for malaria control 2012-2014, at http://www.who.int/malaria/publications/world-malaria-report-2015/en/)

Country	Population (2015)	Total GDP 2012-2014 (USD)	Total malaria funds 2012-2014 (USD)	Average per capita (USD)
Nigeria	185,578,000	1,665,365,877,787	651,741,493	1.17
Democratic Republic of the Congo	77,066,000	98,689,122,034	525,235,539	2.27
South Sudan	12,358,000	40,737,134,236	306,610,362	8.27
Kenya	46,034,000	171,209,650,947	284,747,325	2.06
Uganda	39,014,000	79,173,829,488	277,121,578	2.37
Ethiopia	99,215,000	152,971,928,220	265,831,951	0.89
Côte d'Ivoire	23,092,000	98,281,648,192	249,860,119	3.61
Ghana	27,271,000	132,987,589,514	235,314,456	2.88
Cameroon	23,345,000	94,830,482,893	190,572,589	2.72
Zambia	16,187,000	86,176,547,649	150,384,345	3.10
Madagascar	24,165,000	32,818,301,997	128,891,530	1.78
Mozambique	27,332,000	49,069,798,560	126,454,789	1.54
Angola	24,832,000	350,433,626,318	110,927,295	1.49
Mali	17,769,000	42,299,577,016	105,545,476	1.98
Senegal	14,857,000	46,617,718,771	88,059,637	1.98
Malawi	17,186,000	18,254,102,685	86,721,342	1.68
Liberia	4,540,000	6,027,411,111	76,128,148	5.59
Zimbabwe	15,375,000	46,037,106,109	71,332,469	1.55
Burkina Faso	17,968,000	37,337,762,668	70,600,257	1.31
Burundi	11,048,000	9,510,009,132	66,538,243	2.01
Benin	10,852,000	29,266,808,000	66,475,999	2.04
Sierra Leone	6,386,000	12,651,666,044	57,366,632	2.99
Guinea	12,644,000	20,276,141,235	56,350,249	1.49
Chad	14,055,000	41,960,908,608	33,760,312	0.80
Niger	20,058,000	24,837,023,814	28,040,697	0.47
Central African Republic	4,831,000	5,802,161,934	16,105,877	1.11
Gambia	1,971,000	2,828,801,981	15,879,580	2.69
TOTAL			**4,342,598,289**	

Conclusions

The emergence, expansion or reduction and elimination of VBDs are dynamic processes influenced by social and environmental factors. Some diseases are well known, others less so; some, like yellow fever or plague can be quite easily controlled with tools such as vaccines or antibiotics, while others still pose a difficult challenge to public health systems. Currently, the highest uncontrolled burden is due to arboviruses, with a few viruses representing the majority of severe cases – i.e. dengue viruses, chikungunya virus and Zika virus. Since no simple and efficient tools are available against these diseases, the best option is to mobilize all resources and energies including community participation. The main challenge is the sustainability of action since, while communities can take all the necessary action to eliminate the breeding sites of the main *Ae. aegypti* vector following strong communication and education campaigns, their efforts may not be sustainable since they also have to contend with other primary problems including water, food, work and others. For the poorest and most neglected populations, which have the highest VBD burden (*Figure 3*), the struggle is for survival, and action is taken against mosquitoes only when the disturbance (nuisance or disease) is unbearable – and the same is true for governments. Consequently, the governments of Low and Middle Income Countries (LMIC) strongly engage in vector control plans only when an outbreak is ongoing, which is too late.

The fight against VBDs is correlated to economic development and the improvement of living conditions. Progress in our knowledge of VBDs and new technologies to control vectors will help reduce the burden of these diseases only if the social context is favorable. The successful interventions deployed in the past using military-like approaches were not sustained because they were imposed and unethical. Future approaches must be developed, accepted and supported by the affected communities if sustainability is the target, and such approaches must therefore strive not only for the absence of disease but also for the best possible living conditions.

References

Achee N.L., Youngblood L., Bangs M.J., Lavery J.V. and James S. (2015). "Considerations for the Use of Human Participants in Vector Biology Research: A Tool for Investigators and Regulators", *Vector Borne Zoonotic Dis.*, 15(2), 89-102. doi:10.1089/vbz.2014.1628

Andraud M., Hens N., Marais C. and Beutels P. (2012). "Dynamic Epidemiological Models for Dengue Transmission: A Systematic Review of Structural Approaches", *PLoS ONE*, 7(11), e49085.

Bowman L.R., Runge-Ranzinger S. and McCall P.J. (2014). "Assessing the relationship between vector indices and dengue transmission: a systematic review of the evidence", *PLoS Negl Trop Dis.*, 8(5), e2848. doi:10.1371/journal.pntd.0002848

Dobson A.P. and Carper E.R. (1996). "Infectious diseases and human population history", *Bioscience*, 46(2), 115-126.

Cibulskis R.E., Alonso P., Aponte J., Aregawi M., Barrette A., Bergeron L., Fergus C.A., Knox T., Lynch M., Patouillard E., Schwarte S., Stewart S. and Williams R. (2016). "Malaria: Global progress 2000-2015 and future challenges", *Infect Dis Poverty*, 5(1), 61. doi:10.1186/s40249-016-0151-8

Flacio E., Engeler L., Tonolla M., Lüthy P. and Patocchi N. (2015). "Strategies of a thirteen year surveillance programme on *Aedes albopictus* (Stegomyia albopicta) in southern Switzerland", *Parasites and Vectors*, 8:208. doi:10.1186/s13071-015-0793-6

Floch H. (1954). "Antimalarial campaign in French Guiana", *Bulletin of the World Health Organization*, 11(4-5), 579.

Fouque F., Baumgärtner J. and Delucchi V. (1998). « La démoustication de la plaine de Magadino. II. Écologie des populations du moustique *Aedes vexans* (MEIGEN) et propositions de lutte intégrée », *Bull. Soc. Ent. Suisse*, 71, 439-447.

Goindin D., Delannay C., Ramdini C., Gustave J. and Fouque F. (2015). "Parity and longevity of *Aedes aegypti* according to temperatures in controlled conditions and consequences on dengue transmission risks", *PLoS ONE*, 10(8), e0135489. doi:10.1371/journal.pone.0135489

Huldén L., Huldén H. and Heliövaara K. (2005). "Endemic malaria: an 'indoor' disease in northern Europe. Historical data analysed", *Malaria Journal*, 54:19. https://doi.org/10.1186/1475-2875-4-19.

Lengeler C. (2000). "Insecticide-treated bednets and curtains for preventing malaria", *Cochrane Database Syst Rev.*, (2):CD000363.

Mendis K., Rietveld A., Warsame M., Bosman A., Greenwood B. and Wernsdorfer W.H. (2009). "From malaria control to eradication: The WHO perspective", *Trop Med Int Health*, 14(7), 802-809. doi:10.1111/j.1365-3156.2009.02287.x.

NEGHINA R., NEGHINA A.M., MARINCU I. and IACOBICIU I. (2010). "Malaria, a Journey in Time: In Search of the Lost Myths and Forgotten Stories", *The American Journal of the Medical Sciences*, 340(6), 492-498. Online: https://doi.org/10.1097/MAJ.0b013e31 81e7fe6c.

PATTERSON J., SAMMON M. and GARG M. (2016). "Dengue, Zika and Chikungunya: Emerging Arboviruses in the New World", *West J Emerg Med.*, 17(6), 671-679.

POWELL J.R. and TABACHNICK W.J. (2013). "History of domestication and spread of *Aedes aegypti* – A Review", *Mem Inst Oswaldo Cruz*, Rio de Janeiro, 108(Suppl. I), 11-17.

SERVICE M.W. (1978). "A short history of early medical entomology", *J Med Entomol*, 14(6), 603-626.

World Health Organization (2016). *World Malaria Report 2016*, ISBN: 978 92 4 151171 1.

WILSON A.J., MORGAN E.R., BOOTH M., NORMAN R., PERKINS S.E., HAUFFE H.C., MIDEO N., ANTONOVICS J., MCCALLUM H. and FENTON A. (2017). "What is a vector?", *Philosophical Transactions of the Royal Society B Biological Sciences*, 372(1719). Online: http://dx.doi.org/10.1098/rstb.2016.0085.

Social dimensions factored into mosquito-borne diseases, their prevention and control. From a global perspective to the Zika case in Latin America

Mariam Otmani del Barrio

Vectors, Environment and Society Unit, Special Programme for Research and Training in Tropical Diseases (TDR), World Health Organization (WHO), Geneva, Switzerland

More than half of the world's population is at risk of vector-borne diseases (VBDs). The parasites, bacteria and viruses transmitted to humans by mosquitoes, ticks, snails, flies and other small organisms account for 17% of the global burden of illness and disability (WHO, 2014a; WHO, 2017a); each year, more than a billion people are infected by VBDs and death estimates account for over one million people globally (WHO, 2014a; Cuervo-Parra *et al.*, 2016; WHO, 2004). Mosquito-borne diseases transmitted to humans include parasitic protozoan diseases like malaria, filarial diseases such as lymphatic filariasis (elephantiasis), and viral diseases such as dengue, encephalitis, West Nile fever, yellow fever, chikungunya and Zika virus. The primary mosquito vector species involved in human disease include the *Culex* (Japanese encephalitis, West Nile virus, lymphatic filariasis), *Anopheles* (malaria, lymphatic filariasis in rural areas), and *Aedes* genera (Zika virus [ZIKV], dengue virus, chikungunya virus [CHKV], yellow fever) which can lead to major global health challenges.

Multiple factors combined affect the health of individuals and communities. The socio-economic and environmental conditions in which people are born, live, work and age affect human lives and determine to a large extent people's risk of contracting illness and their ability to prevent and treat diseases, including vector-borne ones.

Political, social and market economy forces and associated inequities in power, money and resources also shape these conditions in which people live over the course of their lives (WHO, 2008).

Inequalities in health associated with different exposures and vulnerabilities to conditions that affect an individual's health are therefore influenced by several, multifaceted factors including age, gender, ethnicity and geographic location (NICE, 2012; IDeA, 2010; IHE and IHI, 2016) but they are also linked to opportunities for education and employment, the physical environment, income levels, social and economic security, lifestyle choices and habits, and experiences of inequality (WHO, 2015). Globally, the health inequalities reflected by the social gradient phenomenon appear across a broad socio-economic spectrum, not just in low income countries but also in middle and high income countries. Those who live at the very end of the inequality scale face social determinants that negatively impact their health, with disadvantaged and vulnerable groups more likely to suffer from a range of human development issues, from poverty and poor housing conditions, to low school attendance and reduced access to health care. Similarly, and although the health outcomes and ecology of mosquito-borne diseases are diverse, the poorest segments of society and resource constraint settings (both financially and in terms of human resources) are disproportionately affected – particularly in least-developed countries with tropical climates. Communities are affected in urban, peri-urban and rural settings but such diseases flourish predominantly among those that face poor living conditions and particularly a lack of access to adequate housing, safe drinking water and sanitation, thus reflecting large structural inequalities in accessing health services and the fundamental determinants of health. Malnourished individuals and those with weakened immunity and inadequate access to health care are especially vulnerable. In low and middle income countries particularly, these diseases also work to exacerbate the cycle of poverty by reducing productivity, the ability to produce food or earn an income and perpetuating socio-economic exclusion (Faburay, 2015). Illness and disability prevent people from working and supporting themselves and their family, causing further hardship to already vulnerable communities or individuals (WHO, 2014a).

There are also important social and demographic inequalities in both exposure to and negative health outcomes that derive from adverse environmental conditions, between countries, within countries and within communities themselves. The risk for human exposure to mosquito-borne diseases is variable in space and time, and is influenced by increased human mobility and displacement, population growth, unplanned urbanization, inadequate water and sanitation, global trade, and environmental changes. The incidence and geographical distribution of mosquito-borne diseases are swayed by a complex interaction of social, demographic, environmental and climatic factors, whose variable impacts affect pathogen transmission patterns, resulting in an intensification, geographical spread, re-emergence or extension of transmission seasons (WHO, 2017b).

Despite decades of interventions to fight mosquito-borne diseases, and success stories notably with malaria control programmes and efforts to combat lymphatic filariasis for example, the burden of mosquito-borne diseases remains a public health concern with uneven progress made across regions and diseases (Lancet editorial, 2017; WHO, 2016; WHO, 2010). The broader socio-economic and environmental contexts therefore have an important influence on how mosquito control mechanisms are adopted by the targeted populations and local health systems.

Multiple vulnerabilities, local disease dynamics: community engagement and socio-ecological approaches for the prevention and control of mosquito-borne diseases

Since there is no single vulnerability but many and the dynamics of disease tend to be localized, evidence shows that community-centred strategies, where not only participation but active engagement and ownership from communities and local authorities is ensured, lead to effective approaches to combat mosquito-borne diseases (Machaca *et al.*, 2002; Basso *et al.*, 2017). This should not imply that the mere success of an intervention relies on communities leading the interventions themselves, but rather it acknowledges the dynamics and social connections that are created by the different actors in the response to mosquito-borne diseases. In addition, studies show that the implementation of mosquito control measures with community involvement can take place at a reasonable cost, yielding promising health outcomes and contributing to the effectiveness, sustainability and community acceptability of the interventions (Alfonso-Sierra *et al.*, 2016; Basso *et al.*, 2017). Recent research in Ecuador shows that even in resource-constrained households people are willing to spend a considerable amount of their income on *Aedes aegypti* control measures (almost 2% of their median monthly income, from liquid sprays and repellents to mosquito coils and bed nets) (Heydari *et al.*, 2017). However, with the arrival of new *Aedes* control technologies (e.g. Wolbachia and genetically modified mosquitoes) it is also urgent that social and ecological considerations and approaches be taken into account, since the effectiveness of these new strategies will largely depend on the perceived benefit-risk balance, efficacy and effectiveness perceptions at the community level, as well as on behavioural and economic barriers to implementation at the community and household levels (Stewart Ibarra *et al.*, 2014; Amin and Hashim, 2014). With the recent rise of chikungunya and Zika, greater attention will be required to understand how communities in vulnerable and resource constraint settings will be able to adapt to these novel interventions and how community engagement strategies will take into account local values, power relations, attitudes, resources and capacities (Musesengwa, 2017).

Community engagement has been increasingly incorporated in global health and control efforts to tackle VBDs; a step further in community engagement, considering the social and cultural realities of the disease (Suarez *et al.*, 2005), acknowledging the preferences of communities and high-risk groups (Grietens *et al.*, 2013), and the continuous need for understanding and incorporating perspectives, risks and vulnerabilities, perceptions and adaptive capacities at the local level is required, particularly under a changing climate, to balance the predominant flow of knowledge and evidence that cascades from the global to the local level in many cases ignoring focalised social perspectives (Feierman *et al.*, 2010; Bardosh, 2014; Bokko *et al.*, 2016). Recent research in the Okavango wetlands shows how community adaptation strategies are informed by perceptions, past memories and historical knowledge of environmental changes and disease outbreaks in the face of climate variability and change, and claims that the knowledge systems of communities contribute to a more comprehensive understanding of the global ecological system in order to inform effective adaptation policies and strategies (Ngwenya *et al.*, 2016). Beyond claiming that social science research has traditionally not given much attention to Neglected Tropical Diseases (NTDs) for example, Bardosh points up the importance of social inquiry and the need to understand factors associated with local policy processes, health systems capacity, community preferences and responses to interventions, education and behaviour change, as well as community participation to effectively implement prevention and control strategies that respond to local needs and local challenges – which, in most cases require addressing not only the mosquito vector or the pathogen, but precisely the fundamental socio-economic and environmental determinants of health influenced by the local cultural, political, social and economic context (Bardosh, 2014; Parker and Allen, 2013; Spiegel *et al.*, 2005). Similarly, a recent study contextualized the emergence and proliferation of the 2012 seasonal dengue outbreak in Pucallpa, the second largest city in the Peruvian Amazon, highlighting the tension between rapidly changing environmental and human settings, including increased travel, rapid urbanization, and inadequate water management, coupled with poor surveillance and underfunded ad-hoc vector control, which contributed to the emergence and proliferation of the dengue outbreak, a public health threat which seems to persist in this region (Charette *et al.*, 2016). This example, in a country with endemic dengue, used a socio-ecological approach to show the importance of regional and municipal governance structures behind the risk of dengue spread in the Amazon basin. It suggests the need to consider contextual social, political, environmental and economic determinants far beyond individual risk factors such as age, sex or immunity status to explain patterns of dengue incidence at the population level and to shape prevention and control efforts in a region with limited local funding that also faces the threat of Zika and chikungunya. Vector-control programs – that require well-resourced, resilient and responsive public health systems – may not always be a priority in national budgets, particularly when facing economic recession or political instability, which pose additional risks to already vulnerable and marginalized communities (Phelan and Gostin, 2016). The resource allocation divide between underfunded endemic regions and the increase in research funding most notably for dengue and malaria (Horstick *et al.*, 2015; Charette *et al.*, 2016; Smith *et al.*, 2014), also suggests a growing concern in non-endemic and developed regions over mosquito-borne diseases such as malaria and dengue, in light of the increasing connectivity and human mobility worldwide (Tatem *et al.*, 2017).

In a global context where trends in primary funding flows on health worldwide appear to be donor-driven, favouring problem-focused vertical initiatives (Clinton and Sridhar, 2017; Dieleman *et al.*, 2016) and presumably perpetuating the biomedical vs. social science dichotomy, effective strategies now more than ever require moving away from siloed mosquito-control efforts, embracing integrated vector management and addressing the structural and systemic challenges of focalized poverty, weak governance and non-resilient health systems. These strategies also require participatory action methods and intersectional thinking into development cooperation practice and research, that not only rethink and challenge established methodologies, practices and interpretations of situations and strategies (Schurr and Segebart, 2012; Bergold and Thomas, 2012), but that also engage with local communities beyond just the "knowledge, attitudes and practices questionnaires" (Bardosh, 2014). Communities must be recognized not as mere distant objects of study and information sources, but as a way to contribute to mosquito-control efforts and improved local health and environmental outcomes by incorporating local knowledge, questioning assumptions and understanding the specific context with its challenges, whether cultural, economic, political or environmental, on an equal footing, and respecting global ethical standards. It is therefore not surprising that communities, even when they are very much aware of the intent of a scientific study, are eager to receive feedback on the research findings, which suggests that it is likely that not all research teams follow due diligence in research implementation, ignoring some of the most basic ethical standards in the process.[1]

Considering gender dynamics and power imbalances in prevention and control efforts to combat mosquito-borne diseases

Gender dynamics interact with the social, economic and biological factors that shape the impact of mosquito-borne diseases and their different health outcomes across and within communities and individuals with different gender identity, sex, ethnicity and age.

A significant amount of research has been conducted in the field of gender and health, with substantial attention given to gender differentials in vulnerability to, and the impact of, particular health conditions

1 Personal communication. July 2014, Maasai Steppe's Simanjiro, Tanzania. In the context of a research study visit, a Maasai community member expressed himself to the author of this paper during a conversation over a meal organized with community members, as follows: "We appreciate your visit, but please come back and let us know the results of your work, not everyone comes back," alluding to other past experiences with researchers.

(Vlassoff, 2007; Vlassoff and Garcia Moreno, 2002; Rathgeber and Vlassoff, 1993).

Power dynamics and traditional or cultural norms and practices entrenched in gender roles and relationships have a strong influence, whether in seeking or accessing health care equitably, using treatment, or applying prevention strategies to mosquito-borne diseases (Garley *et al.*, 2013; WHO, 2009). It is indeed well acknowledged that socially determined gender norms imply in many cases that women often carry the extra burden of unpaid care work for looking after sick family members, for example – even when they are sick themselves (Stevens, 2016), which results in them incurring high opportunity costs; or that they may be less involved in drying mosquito breeding sites or insecticide spraying activities in some settings, while playing leading roles in the health education of local communities in the context of environmental and vector control campaigns (Hamid *et al.*, 1996). In many malaria-endemic areas, certain activities traditionally assigned to men may increase exposure to malaria vectors, particularly for those working in gold mines or forest logging, or working at night (WHO, 2009; Kvinnoforum, 2006). Moreover, behavioural factors resulting in the disproportionate exposure of men to the infected mosquito vector may have important roles in West Nile virus (WNV) illness in certain regional contexts (Theiler *et al.*, 2008). However, in cultures where women traditionally tend the animal herds, they are more likely to be exposed to Rift Valley Fever (LaBeaud *et al.*, 2007). Some studies highlight that women and children are also more likely to use insecticide treated nets (ITNs) in sub-Saharan malaria endemic areas, suggesting the need to better understand decision-making at the household level when there are not sufficient ITNs to cover all members; while adolescent boys remain the least likely group to use these nets, being less concerned about malaria during the dry season (Birhanu *et al.*, 2015; Garley *et al.*, 2013).

The literature also shows a pattern of male predominance in the reported number of dengue cases among persons 15 years or older in different Asian countries (Anker and Arima, 2011). Understanding gender considerations implies assessing existing differentials in dengue incidence and other mosquito-borne diseases, as well as potential differentials in behaviours, perceptions and experiences by sex, gender and age, which is important because biological factors and gender dimensions can change over the life cycle, as gender-related factors may differ across countries that also have different health outcomes across regions and within regional locations (Anker and Arima, 2011; Prasith *et al.*, 2013).

Less attention has been given, however, to the threat and scourge to human rights that is the universal phenomenon of gender-based violence, and its impact on the health consequences from mosquito-borne diseases.

As Vlassoff points out, socio-economic and biological determinants, among others, and their consequences on human health and illness interact in the case of gender-based violence (Vlassoff, 2007). In India and Bangladesh, for example, factors such as education and economic empowerment play a critical role in deciding whether to seek help when suffering violence, although informal sources of help are used more often than institutional services (Paul, 2016; Parvin *et al.*, 2016). In Tanzania, evidence shows that help-seeking behaviour was facilitated when women solely owned capital assets rather than when these were owned jointly with their partners (Vyas and Mbwambo, 2017).

Evidence shows that female (including transgender women) survivors of violence, whether physical, sexual, emotional or economic, may hide their injuries or abusive situations from others, including from health services, and may not seek support, because of socio-cultural context shame, social stigma or fear of intensification of the violence (WHO, 2014b; Naved *et al.*, 2006). In this case, it is also likely that those who experience gender-based violence and are affected by mosquito-borne diseases may have less access to health care, fewer health-seeking behaviours or less freedom of choice. This violence interacts with the roles and position of different gender identities in the household and in communities, influencing decision-making processes and risks of infection.

The recent emergence of Zika virus in the Americas, a region with very high levels of sexual violence against women, revealed the ongoing intensification of environmental and social factors that have fuelled arbovirus transmission (Ali *et al.*, 2017). Economic and gender inequality, discrimination and poverty influence the risks and impacts of Zika, which disproportionately affect women, that in most cases do not exercise their sexual and reproductive health rights. Evidence shows that certain recommendations by authorities, such as pregnancy postponement, presented limited success likely due to limited access to contraceptive methods, low or lack of sexual education, and high rates of sexual assault (Sedgh *et al.*, 2014; WHO, 2012) which, coupled with barriers to safe abortion, further exacerbated the impact of Zika virus on women and particularly on the poor (Sandy, 2016; Cabezas, 2016; Strochli and Cohen, 2016). A recent analysis conducted in Cúcuta (Colombia), showed that new-borns and women (in the range of 15 to 39 years old) were at the highest risk of being infected by Zika (Anaya *et al.*, 2017); this study also highlighted that low socio-economic status and a high load of previous infection were the primary risk factors associated with the development of Guillain-Barré syndrome (GBS) in people infected with Zika virus; patients with lower socio-economic status also tend to live closer to areas where waste water contamination occurs, which reinforces the need to address socio-economic and environmental determinants of health and to identify inequalities through comprehensive global health efforts to fight viral diseases transmitted by *Aedes aegypti* mosquitoes. In Rio de Janeiro, even discounting pregnant women, a massive increase of Zika incidence (almost 90% higher) in women aged 15 to 65 compared to men was documented for the years 2015-2016 (Coelho *et al.*, 2016), which suggests that given that not only vector transmission but also sexual transmission occurs, women are doubly confronted with the challenges associated with traditionally assigned roles in society (e.g. in certain contexts it is more likely that women stay in the household and increase exposure to the vector), and with the challenges associated with the exercise of their sexual and reproductive rights, unequal power relations within asymmetrical sexual relationships (Gonzalez Vélez and Diniz, 2016), and ultimately the impact of inequalities on women's physical and mental health.

References

Alfonso-Sierra E., Basso C., Beltrán-Ayala E., Mitchell-Foster K., Quintero J., Cortés S., Manrique-Saide P., Guillermo-May G., Caprara A., de Lima E. and Kroeger A. (2016). "Innovative dengue vector control interventions in Latin America: what do they cost?", *Pathogens and Global Health*, 110(1), 15-24.

Ali S., Gugliemini O., Harber S., Harrison A., Houle L., Ivory J., Kersten S., Khan R., Kim J., LeBoa C. *et al.* (2017). "Environmental and Social Change Drive the Explosive Emergence of Zika Virus in the Americas", *PLoS Negl Trop Dis*, 11(2).

Amin L. and Hashim H. (2014). "Factors influencing stakeholders attitudes toward genetically modified Aedes mosquito", *Science and Engineering Ethics*, 21(3), 655-681.

Anaya J.M., Rodriguez Y., Monsalve D.M., Vega D., Ojeda E., Gonzalez-Bravo D., Rodriguez-Jiménez M., Pinto-Diaz C.A. *et al.* (2017). "A comprehensive analysis and immunobiology of autoimmune neurological syndromes during the Zika virus outbreak in Cúcuta. Colombia", *Journal of Autoimmunity*, 77, 123-138.

Anker M. and Arima Y. (2011). "Male-female differences in the number of reported incident dengue fever cases in six Asian countries", *Western Pacific Surveillance and Response Journal*, 2(2), 17-23.

Bardosh K. (2014). "Global aspirations, local realities: the role of social science research in controlling neglected tropical diseases", *Infectious Diseases of Poverty*, 3(35), 1-15.

Basso C., Garcia da Rosa E., Lairihoy R., Caffera R.M., Roche I., González C., da Rosa R., Gularte A., Alfonso-Sierra E., Petzold M., Kroeger A. and Sommerfeld J. (2017). "Scaling up of an Innovative Intervention to Reduce Risk of Dengue, Chikungunya and Zika transmission in Uruguay in the Framework of an Intersectoral Approach with and without Community Participation", *American Journal of Tropical Medicine and Hygiene*. Available from: https://doi.org/10.4269/ajtmh.17-0061.

Bergold J. and Thomas S. (2012). "Participatory Research Methods: A Methodological Approach in Motion", *Forum Qualitative Sozialforschung / Forum: Qualitative Social Research*, 13(1), Art. 30. Available from: http://nbn-resolving.de/urn:nbn:de:0114-fqs1201302.

Birhanu Z., Abebe L., Sudhakar M. and Dissanayake G. (2015). "Access to and use gaps of insecticide-treated nets among communities in Jimma Zone, southwestern Ethiopia: baseline results from malaria education interventions", *BMC Public Health*, 15(1304), 29 December.

Bokko A., Cisse G., Kone B. and Dedy S. (2016). « Croyances locales et stratégies d'adaptation aux variations climatiques à Korhogo (Côte d'Ivoire) ». *Tropicultura*, 34(1), 40-46.

Cabezas J. (2016). "El Salvador urges against pregnancies until 2018 as Zika virus spreads". Available from: http://www.reuters.com/article/us-health-zika-el-salvadoridUSKCN0UZ2SP.

Charette M., Berrang-Ford L., Llanos-Cuentas E.A., Cárcamo C. and Kulkarni M. (2016). "What caused the 2012 dengue outbreak in Pucallpa, Peru? A socio-ecological autopsy", *Social Science and Medicine*, 174, 122-132.

Clinton C. and Sridhar D. (2017). "Who pays for cooperation in global health? A comparative analysis of WHO, the World Bank, the Global Fund to Fight HIV/AIDS, Tuberculosis and Malaria, and Gavi, the Vaccine Alliance", *The Lancet*, 390(10091), 27 January, 324-332.

Coelho F., Durovni B., Saraceni V., Lemos C., Codeco C.T., Camargo S. *et al.* (2016). "Higher incidence of Zika in adult women than adult men in Rio de Janeiro suggests a significant contribution of sexual transmission from men to women", *International Journal of Infectious Diseases*, 51(October), 128-132.

Cuervo-Parra J.A., Romero Cortés T. and Ramirez-Lepe M. (2016). "Mosquito-Borne Diseases, Pesticides Used for Mosquito Control, and Development of Resistance to Insecticides. Insecticides Resistance", Prof. Stanislav Trdan (Ed.), *InTechOpen*, doi:10.5772/61510. Available from: https://www.intechopen.com/books/insecticides-resistance/mosquito-borne-diseases-pesticides-used-for-mosquito-control-and-development-of-resistance-to-insect.

Dieleman J., Schneider M., Haakenstad A., Singh L., Sadat N., Birger M., Reynolds A., Templin T., Hamavid H., Chapin A. and Murray C.J. (2016). "Development assistance for health: past trends, associations, and the future of international financial flows for health". *The Lancet*, 387(10037), 18 June, 2536-2544.

Faburay B. (2015). "The case for a 'one health' approach to combating vector-borne diseases", *Infection Ecology and Epidemiology*, 5(1), Article 28132.

Feierman S., Kleinman A., Stewart K., Farmer P. and Das V. (2010). "Anthropology, knowledge flows and global health", *Global Public Health*, 5(2), 122-128.

Garley A.E., Ivanovich E., Eckert E., Negroustoueva S. and Ye Y. (2013). "Gender differences in the use of insecticide-treated nets after a universal free distribution campaign in Kano State, Nigeria: post-campaign survey results", *Malaria Journal*, 12(119), 10 April.

Gonzalez Vélez A.C. and Diniz S.G. (2016). "Inequality, Zika epidemics, and the lack of reproductive rights in latin America". *Reproductive Health Matters*, 24(48), 57-61.

Grietens K.P., Muela Ribera J., Soto V., Tenorio A., Hoibak S., Rosas Aguirre A., Toomer E., Rodriguez H., Llanos Cuentas A., D'Alessandro U., Gamboa D. and Erhart A. (2013). "Traditional Nets Interfere with the Uptake of Long-Lasting Insecticidal Nets in the Peruvian Amazon: The Relevance of Net Preference for Achieving High Coverage and Use", *PLoS ONE*, 8(1), 2 January.

Hamid S.R., Mohamedani A.A., Mirgani E.M. and Mahgoub I.A. (1996). "Gender aspects and women's participation in the control and management of malaria in Central Sudan", *Social Science & Medicine*, 42(10), 1433-1446.

Heydari N., Larsen D.A., Neira M., Ayala E.B. and Fernandez P. (2017). "Household Dengue Prevention Interventions, Expenditures, and Barriers to *Aedes aegipti* Control in Machala. Ecuador", *Int. J. Environ. Res. Public Health*, 14(2).

Horstick O., Tozan Y. and Wilder-Smith A. (2015). "Reviewing Dengue: Still a Neglected", *PLoS Negl Trop Dis*, 9(4), 30 April.

Improvement and Development Agency (IDeA) (2010). *The social determinants of health and the role of the local government.* London, F. Campbell.

Kvinnoforum and The Roll Back Malaria Partnership Secretariat (2006). *A guide to gender and malaria resources.* Stockholm, Kvinnoforum.

LaBeaud A.D., Ochiai Y., Peters C.J., Muchiri E.M. and King C.H. (2007). "Spectrum of rift valley fever virus transmission in Kenya: Insights from three distinct regions", *Am J Trop Med Hyg*, 76(5), 795-800.

Lancet Editorial (2017). "Precision global health: beyond prevention and control", *Lancet Global Health*, 5(e1), January.

Machaca J., Llontop F., Pasapera F. and Castañeda C. (2002). «Eliminación mecánica de huevos del Aedes aegypti para la erradicación del dengue urbano. Localidad de Sechura-Piura, Peru, Abril-December 2001», *Revista Peruana de Epidemiología*, 10(1).

Musesengwa R., Chimbari M. and Mukaratirwa S. (2017). "Initiating community engagement in an ecohealth research project in Southern Africa", *Infect Dis Poverty*, 6(22), 7 March.

National Institute for Health and Care Excellence (NICE) (2012). *Health inequalities and population health*. NICE, UK.

Naved R., Azim S., Bhuiya A. and Persson L.A. (2006). "Physical violence by husbands: magnitude, disclosure and help-seeking behaviour of women in Bangladesh", *Soc Sci Med*, 62(12), 2917-2929.

Ngwenya B.N., Thadaku O.T., Magole L. and Chimbari M.J. (2016). "Memories of environmental change and local adaptations among molapo farming communities in the Okavango Delta, Botswana: a gender perspective", *Acta Tropica*, 19 November.

Parker M. and Allen T. (2013). "De-politicizing parasites: reflections on attempts to control the control of neglected tropical diseases", *Med Anthropol*, 33(3), 223-239.

Parvin K., Sultana N. and Naved R.T. (2016). "Disclosure and help seeking behavior of women exposed to physical spousal violence in Dhaka slums", *BMC Oublic Health*, 16(383). doi:10.1186/s12889-016-3060-7

Paul S. (2016). "Intimate partner violence and women's help-seeking behavior: Evidence from India", *Journal of Interdisciplinary Economics*, 28(1), 53-82.

Phelan A. and Gostin L. (2016). "The human Rights Dimensions of Zika", Available from: https://academicimpact.un.org/content/human-rights-dimensions-zika

Prasith N., Keosavanh O., Phengxay M., Stone S., Lewis H.C., Tsuyuoka R., Matsui T., Phongmanay P., Khamphaphongphaneb B. and Arima Y. (2013). "Assessment of gender distribution in dengue surveillance data, the Lao People's Democratic Republic", *Western Pacific Surveillance and Response Journal*, 4(2), 17-24. doi:10.5365/wpsar.2012.3.4.020

Rathgeber E. and Vlassoff C. (1993). "Gender and tropical diseases: a new research focus", *Soc Sci Med*, 37(4), 513-520.

Roberto Suarez M., Olarte M.F., Ana M.F.A. and Catalina Gonzalez U. (2005). "Is what I have just a cold or is it dengue? Addressing the gap between the politics of dengue control and daily life in Villavicencio, Colombia", *Soc Sci Med*, 61(2), 495-502.

Sandy M. (2016). "Brazilian Legislators Look to Increase Abortion Penalties in the Wake of Zika Outbreak". Available from: http://time.com/4230975/brazil-abortion-laws-zikaoutbreak.

SCHURR C. and SEGEBART D. (2012). "Engaging with feminist postcolonial concerns through participatory action research and intersectionality", *Geogr. Helv.*, 67, 147-154.

SEDGH G., SINGH S. and HUSSAIN R. (2014). "Intended and Unintended Pregnancies Worldwide in 2012 and Recent Trends", *Stud Fam Plann*, 45(3), 301-314.

SMITH D., PERKINS T., REINER R.J., BARKER C., NIU T., CHAVES L., ELLIS A., GEORGE D., LE MENACH A., PULLIAM J., BISANZIO D., BUCKEE C. and CHIYAKA C. (2014). "Recasting the theory of mosquito-borne pathogen transmission dynamics and control", *Trans R Soc Trop Med Hyg*, 108(4), 185-197.

SPIEGEL J., BENNETT S., HATTERSLEY L., HAYDEN M., KITTAYAPONG P., NALIM S. and WANG D. (2005). "Barriers and bridges to prevention and control of dengue: the need for a social-ecological approach", *Ecohealth*, 2(4), 273-290.

STEVENS E. (2016). "In Peru, local women take aim at dengue", *Oxfam Impact*, October.

STEWART IBARRA A., LUZADIS V., CORDOVA M., SILVA M., ORDOÑEZ T., AYALA E. and RYAN S. (2014). "A social-ecological analysis of community perceptions of dengue fever and Aedes aegypti in Machala, Ecuador", *BMC Public Health*, 14(1135).

STROCHLIC N. and COHEN N.S. (2016). "Where Women with Zika Fear Prison. In El Salvador, abortion–suspected or real–could carry a 50-year sentence". Available from: http://news.nationalgeographic.com/2016/10/zika-el-salvador-women/

TATEM A.J., JIA P., ORDANOVICH D., FALKNER M., HUANG Z., HOWES R., HAY S.I., GETHING P.W. and SMITH D.L. (2017). "The geography of imported malaria to non-endemic countries: a meta-analysis of nationally reported statistics", *The Lancet Infectious Diseases*, 17(1), 98-107.

The UCL Institute of Health Equity (IHE) and Ifakara Health Institute (IHI) (2016). "Creating healthy cities in Tanzania. Using the social determinants framework to understand how urbanization impacts on human development in Tanzanian cities", *The UK Department for International Development* (DFID).

THEILER R.N., RASMUSSEN S.A., TREADWELL T.A. and JAMIESON D.J. (2008). "Emerging and Zoonotic Infections In women", *Infect Dis Clin North Am*, 22(4), 755-viii.

Vlassoff C. (2007). "Gender Differences in Determinants and Consequences of Health and Illness", *J Health Popul Nutr*, 25(1), 47-61.

Vlassoff C. and Garcia Moreno C. (2002). "Placing gender at the centre of health programming: challenges and limitations", *Soc Sci Med*, 54(11), 1713-1723.

Vyas S. and Mbwambo J. (2017). "Physical partner violence, women's economic status and help-seeking behavior in Dar es Salaam and Mbeya, Tanzania". *Global Health Action*, 10(1), 1290426.

World Health Organization (2004). *The world health report 2004 – changing history*. Geneva, WHO.

World Health Organization (2014a). *A global brief on vector-borne diseases.* Geneva, WHO.

World Health Organization (2008). *WHO Commission on Social Determinants of Health, Closing the Gap in a Generation: Health Equity through Action on the Social Determinants of Health*. Commission on Social Determinants of Health Final Report. Geneva, WHO.

World Health Organization (2009). *Women and Health: Today's Evidence Tomorrow's Agenda*. Geneva, WHO.

World Health Organization (2010). *Working to overcome the global impact of neglected tropical diseases: first WHO report on neglected tropical diseases.* Geneva, WHO.

World Health Organization (2012). *Understanding and Addressing Violence Against Women: Sexual Violence*. Geneva, WHO.

World Health Organization (2014b). *Consolidated guidelines on HIV prevention, diagnosis, treatment and care for key populations*. Geneva, WHO, reprint 2016.

World Health Organization (2015). *Intergenerational equity briefing. Review of social determinants of health and the health divide in the WHO European Region*. Copenhagen, WHO Regional Office for Europe.

World Health Organization (2016). *World Malaria Report 2016*. Geneva, WHO.

World Health Organization (2017a). *Inheriting a sustainable world? Atlas on children's health and the environment*. Geneva, WHO.

World Health Organization (2017b). *Global Vector Control Response 2017-2030. Background document to inform deliberations during the 70th session of the World Health Assembly*. Geneva, WHO.

Shifting socio-ecologies of dengue fever in the United States: Lessons from Florida, Texas, and Arizona

Melinda BUTTERWORTH

Assistant Professor, Willamette University, Salem

Dengue fever is a viral disease spread by *Aedes aegypti* and *Ae. albopictus* mosquitoes. While many cases are asymptomatic, common symptoms include fever, joint pain, and rash. In the most severe cases, hemorrhaging and death can occur. High incidence of the disease occurs in tropical areas of Latin America and Southeast Asia where it has been a concern for centuries. However, there has also been local transmission in southern regions of Europe and North America, as well as in northern Australia. In the case of the United States, dengue fever is a re-emerging disease, as outbreaks occurred in the 1800s and early 1900s. Collectively, dengue is an immensely important vector-borne disease, as nearly 400 million people may be infected with the virus each year (Bhatt *et al.*, 2013).

The case of dengue presents a fascinating example of a complex socio-ecological problem. Its re-emergence is as much a result of social processes, such as the movement of people, limitations of public health capacities, and the intense burden of poverty as it is of ecological factors like changing viral activity, mosquito abundance patterns, and ultimately climate change. These dynamics challenge static notions of dengue's geography, forcing us to grapple with its changing epidemiological patterns (Guha-Sapir and Schimmer 2005). Such is the case in the United States, where epidemic dengue has returned as a public health concern after decades without local transmission. The recent US outbreaks are set amidst a growing trend of increased global dengue incidence and distribution, and require an integrated effort amongst local residents and numerous agencies tasked with managing viruses, vectors, and ecological habitats.

At the same time, these outbreaks challenge predominant discourses that suggest dengue is a disease limited to poor locations and "unclean environments" (Mulligan *et al.*, 2012a). Given this, it is important to recognize the social and ecological conditions present in the United States that can and do support dengue fever transmission.

Dengue's association with poverty in the Global South is attributed to a lack of sanitation, public health infrastructure, and high population density in urban areas, coupled with suitable climates supporting transmission (WHO, 2017). Research in Brazil, which has the highest dengue burden in the Americas (PAHO, 2017), demonstrates that the poor are hit particularly hard by the disease (Braga *et al.*, 2010). But importantly, dengue is not *limited* to these areas, a "paradox" described by Teixeira *et al.* (2009) where dengue risk is present in rich and poor communities, with different risk factors associated with varying population groups (Braga *et al.*, 2010).

Some scholars question the paucity of empirical evidence to support explicit claims linking dengue to poverty (Mulligan *et al.*, 2015), as recent research and news articles highlight dengue fever cases outside of poor communities (Teixeira *et al.*, 2002, 2009; Gale and Srivastava, 2010; Chandra, 2012; Fazlulhaq, 2012), including authochonous transmission in the Global North. The challenges posed by dengue's appearance in non-poor and non-tropical contexts are multiple. First, it can cause the disease and vector habitats to be overlooked by planners and health workers in these areas (Mulligan *et al.*, 2012b; Alley and Sommerfield, 2014). Second, the association of certain peoples and places with various diseases has acted to stigmatize and place blame on marginalized groups thought to present a health risk to others (Craddock and Brown, 2009). In both regards, it is important to recognize the social and ecological conditions present *within* the United States that can support and transmit dengue fever, and to avoid treating dengue outbreaks as simply "emanating"[2] from elsewhere. This of course does not discredit the strong linkages between structural inequality, poverty, and disease, but it does suggest these new locations are not aberrations, but represent a broader geography of dengue requiring investigation (Guha-Sapir and Schimmer, 2005).

---

2 Alley and Sommerfield (2014, 87) noted in Brazil that, "local perceptions that dengue outbreaks usually emanated from favelas," despite disease risk outside of favelas. I draw on this regarding possible perceptions of US dengue outbreaks "spilling over" from Central America.

This chapter will address the re-emergence of dengue fever in the United States, by first presenting a brief overview of mosquito-borne disease and control, including dengue, and its associated socio-ecological factors. Following this, three distinct regions of the southern US, each with its own history of dengue fever, management strategies for *Ae. aegypti* (the focus of this chapter because it is the prominent vector), and ecological characteristics are examined. The first is southern Florida, which is predominated by tropical savanna and monsoonal climates,[3] and experienced a recent dengue outbreak in 2009/2010, and has had localized authochonous transmission in the state since. The second is southern Texas, with a humid-subtropical climate where transmission has occurred along the Texas-Mexico border, and potentially undetected in the city of Houston. The third is southern Arizona; a hot to semi-arid region located in the Sonoran Desert. While no locally acquired cases have been confirmed, *Ae. aegypti* is well established and the bordering state of Sonora, Mexico has active dengue transmission.

After introducing the history and context of each of these three sites, this chapter addresses the myriad of social, political, and ecological issues inherent in the re-appearance of dengue fever in the southern US. These include local public health and vector control surveillance and response capacities, community vulnerability, and climate change.

The historical burden of mosquito-borne diseases in the US

Home to 174 species of mosquitoes (Darsie and Ward, 2005), the insect has a long history of plaguing early European settlers and Native American populations, and as Patterson 2009 (p. 6) writes, "At the beginning of the twentieth century, most Americans considered mosquitoes, like death and taxes, an unavoidable part of the human condition." West Nile virus is the most common mosquito-borne disease reported in the US today. A total of 43,937 cases have been reported to the Centers for Disease Control and Prevention (CDC) since the virus' arrival in 1999 through 2015 (CDC, 2016a). Other endemic mosquito-borne diseases in the US include La Crosse encephalitis, Eastern Equine encephalitis, and St. Louis encephalitis. While these diseases can produce severe consequences in a small number of individuals, the overall burden

[3] The climatic descriptions are based on the Köppen climate classification.

of mosquito-borne diseases today is significantly less than that in the 1700s through the early 1900s. During this past era, large epidemics of serious and at times fatal mosquito-borne diseases plagued much of the United States.

Malaria was found throughout much of the midwestern, southern, and eastern US in the late 1800s. By the early decades of the 1900s, this malaria-stricken region had largely been condensed to the southeastern portion of the US (CDC, 2016b). Sickness and death continued to drop, from 400 cases and 60 deaths per 1,000,000 individuals in 1920 to just under 30 cases and 2 deaths per 1,000,000 individuals in 1946. The US was considered to be malaria-free shortly thereafter, owing to the use of DDT and the eradication of breeding habitats (CDC, 2016b).

Unlike malaria, dengue and yellow fever viruses were not endemic in the United States, yet they still presented a heavy disease burden. As many as 150,000 deaths may have occurred from periodic yellow fever outbreaks in the 1700 and 1800s (Patterson, 1992). Likewise, dengue fever caused particularly large outbreaks in 1873, 1875, and 1922 (Beaumier *et al.*, 2014) with the last documented regional outbreak in 1945 in Louisiana (Hayes *et al.*, 1971). However, the related chikungunya virus may have actually been the agent in some of these cases (Kuno, 1995).

The aforementioned diseases were notable challenges for workers constructing the Panama Canal, and to soldiers at training facilities in the southern US (CDC, 2016b). The presence of disease vectors and nuisance mosquitoes also presented problems in parts of the US that were trying to grow their tourism and real estate sectors (Patterson, 2004, 2009). The dredging and removal of swamps was important for eliminating stagnant water for the mosquito to breed in, while chemical control, such as DDT, was widely used to kill the pest. These programs were indeed successful in finally eliminating malaria transmission within the US. While *Anopheles* mosquitoes that can spread malaria are still found throughout the country today, improvements in settlement locations and infrastructure no longer provide an effective transmission environment.

The health consequences posed by *Ae. aegypti* also prompted large-scale eradication efforts, led by what is now the Pan American Health Organization. As a result, the Americas saw the successful elimination of the mosquito in 18 countries across the Americas by 1962 through the use of intensive top-down programs and chemical insecticides (Anonymous,

1971). However, US participation was limited and ultimately unsuccessful in the elimination of *Ae aegypti*, posing a possible source of re-infestation for the rest of the Americas (Soper, 1963; Anonymous, 1971). Ultimately, the larger success of the eradication programs did not last. Weakening control programs, coupled with urban growth and transportation, aided the mosquito in recovering its lost territory throughout the Americas (Gubler and Clark, 1995; Hotez, 2016).

A resurgence of dengue fever has followed this re-infestation of *Ae. aegypti*. The Pan American Health Organization (PAHO, 2017) reports that a total of 2,338,848 dengue fever cases were reported in the Americas in 2016. With more circulating serotypes (or strains of the disease) and individuals being exposed to subsequent infections, cases of the hemorrhagic form have also been on the rise (PAHO, 2013). Urbanization, population migration, and climate change all stand to further impact the geographic range of *Ae. aegypti* and ultimately dengue fever in the Americas. While theories of epidemiologic transition highlight the decreasing burden of infectious diseases within the Global North (Omran, 1971), we have learned that we are in no way in a post-infectious disease era. While the US remains on the climatic boundaries for *Ae. aeygpti* survival and is comparatively better resourced to manage disease outbreaks than many other countries in the Americas, the dogged persistence of *Ae. aegypti* and the recent outbreaks of dengue within the US suggest the need to take these concerns seriously.

Dengue fever in the United States

Ae. aegypti is the prominent vector of dengue fever. The mosquito bites during the day, prefers to feed on humans, and breeds in artificial and manmade containers that are often found in and around yards. This includes flower saucers, toys, pet bowls, discarded tires, clogged gutters, trash, and in depressions of some plants (*Figure 1*). The eggs can survive desiccation for several months during dry periods, and hatch later once precipitation or water returns (CDC, 2012). In much of the southern US, urban and suburban regions provide ideal habitat for the mosquito to breed in close proximity to humans.

Figure 1. *Examples of manmade and natural breeding sites for Aedes aegypti in urban backyards in Florida. Photos taken by author in summer 2013.*

Believed to be native to Africa, the mosquito likely spread to the Americas via trans-Atlantic ships during the 1500s (Soper, 1967). Ideal temperatures range from 20-28°C, though the mosquito may remain active in temperatures from 15-40°C (Christophers, 1960, 548-551). These temperature parameters are understood to govern the global distribution of *Ae. aegypti* to between the 10°C January and July temperature contours, which approximate to 35°S-45°N latitudes (Christophers, 1960, 35-36), placing much of the southern US within the climatic range for potential seasonal *Ae. aegypti* activity, though not necessarily dengue transmission.

The most recent estimations by the CDC (2017) suggest that *Ae. aegypti* may be present in as many as 29 US states, ranging as far north as Connecticut in the east, and Utah in the west. However, the climatic conditions within the southeastern US are comparatively more suitable for *Ae. aegypti* and dengue transmission (Butterworth *et al.*, 2017), and the potential for transmission exists across a wide range of habitats from tropical South Florida to semi-arid southern Arizona.

The case of Florida

Large outbreaks of dengue fever in Florida are believed to have occurred several times in the 1800s and early 1900s, often linked to transportation via shipping and railroad lines (see Beaumier *et al.*, 2014). In order to develop a tourism and residential industry, the state was forced to aggressively adapt to its mosquito problems through chemical and ecological control methods (Patterson, 2004). While imported cases were periodically reported in Florida, autochonous dengue transmission was absent following the outbreak of 1934 and was not detected again until 2009 in Key West. Since 2010, locally acquired cases have been reported in Miami-Dade, Broward, St. Lucie, Martin, Palm Beach, Osceola, Broward, and Monroe counties (USGS, 2017).

Key West's outbreak began in July of 2009 and lasted approximately 18 months, with the majority of cases occurring during the summer of 2010. A total of 88 locally acquired cases were reported during this time (Florida Department of Health, 2017), however serological evidence suggests that a larger segment (3-5%) of the local population may have actually contracted the virus (Radke *et al.*, 2012). A number of factors make the island of Key West susceptible to such outbreaks. The older portion of the island where the outbreak was centered consists of older homes nestled on small lots, many surrounded by dense vegetation and lacking air conditioning. These factors were linked to dengue cases (Radke *et al.*, 2012) as they, along with old underground cisterns, provide accessible breeding sites and access to blood meals for *Ae. aegypti.* The climate, governed by mild winters and warm, wet summers, further provides ideal conditions for *Ae. aegypti* survival for long periods of the year. From the perspective of stakeholders tasked with managing the outbreak, the proximity of the island to other areas with ongoing transmission further increases outbreak risk (Hayden *et al.*, 2015). The islands of Cuba, Hispaniola, and Puerto Rico have all experienced a rise in epidemic dengue cases, and the cruise ship industry that brings daily visitors to Key West provides an additional opportunity for viral introduction to the island.

The 2009 outbreak was eventually maintained through a combination of aggressive bottom-up and top-down mosquito-control measures, which included door-to-door surveillance and point-source larvaciding, adulticide application, public outreach, and cross-agency collaborations to assist with the control of standing water, known as the Action to Break the Cycle of Dengue (ABCD) campaign (Hayden *et al.*, 2015). Local agencies have considered introducing sterile male mosquitoes as a control strategy, however the plan has been met by mixed opinions by local residents (Ernst *et al.*, 2015). While the ABCD campaign was ultimately successful in halting dengue transmission, *Ae. aegypti* remains established on Key West. Therefore, the risk of a future dengue outbreak remains on the radar of those on the island (Hayden *et al.*, 2015).

Interviews conducted with residents on the island (Butterworth, personal communications, 2012 and 2013) found that several residents eventually diagnosed with dengue were not initially tested for the disease by their physicians. In some cases the patients were told that dengue infection was unlikely, and only through follow-up visits was the diagnosis confirmed. Others were told that it likely was dengue, but since there is no treatment there was no point in obtaining a laboratory confirmed diagnosis at the time. Still other patients only learned of likely dengue infection through the CDC serological survey on the island. Only one case of locally acquired dengue, contracted in 2016, has been reported on the island since the 2009/2010 outbreak.

The case of Texas

The state of Texas has a history of dengue fever that parallels Florida in many ways, as the historical outbreaks in Florida were often part of larger regional outbreaks across the southeastern US that included Texas. Outbreaks from 1885 onwards are believed to be linked to an increase in the state's transportation networks, which impacted Texas cities such as Austin, Galveston, Houston, and Brownsville through 1941 (Ehrenkranz *et al.*, 1971). While transmission was not limited solely to these cities, these locations share a humid subtropical climate, dominated by summers that are hot and humid, and generally mild winters.

After a nearly 40 year absence, several locally-acquired dengue cases were reported in 1980, followed by a handful of outbreaks, predominantly in communities near or on the Texas/Mexico border (Bouri *et al.*, 2012). Bi-national entomologic and epidemiologic surveys have shown that communities on the Mexico side of the border tend to experience a significantly heavier burden of dengue; generally attributed to differences in infrastructure, notably access to air conditioning and window screens on the Texas side (Reiter *et al.*, 2003; Ramos *et al.*, 2008). Nevertheless, Ramos *et al.* (2008, 376) conclude that while differing levels of protective social infrastructure do contribute to a higher prevalence of dengue fever in Mexico that, "despite living conditions at home, that include the availability of air conditioning, substantial numbers of Brownsville residents acquired dengue infection", in addition to overwintering populations in both communities. In 2013, 53 patients were confirmed to have acquired dengue in Cameron, Hildago, and Willacy counties in Texas (Thomas *et al.*, 2016). As with other outbreaks along the southern border, this one was linked to a larger outbreak in northern Mexico (Thomas *et al.*, 2016).

While the burden of dengue is accepted to be higher on the Mexico side of the border, research has shown that the underreporting of cases is a concern in Texas. In Brownsville, Texas, serological tests suggest that close to 40% of residents in the town had previously contracted dengue, which was higher than the numbers reflected in official counts (Brunkard, 2007; Ramos *et al.*, 2008). Murray *et al.* (2013) found evidence of undetected dengue transmission from 2003-2005 in the city of Houston, Texas. This is particularly important because it suggests that dengue transmission can occur undetected for extended periods of time in places, including in the US, perhaps because it is not being looked for. The Houston cases also highlight dengue fever transmission within Texas in areas far removed from the border region. A recent study of physicians in southern Texas further shows a need for increased diagnostic capacities, as it revealed that just over half were able to properly identify dengue symptoms (Adam *et al.*, 2017). This research demonstrates the need for enhanced diagnostic and surveillance measures for dengue fever and recognition that existing disease maps may not be sufficient for accurate diagnoses.

The case of Arizona

The most common disease to plague Arizona is West Nile virus, as the state is home to some of the highest national incidences of reported disease in some years (CDC, 2016). However, *Ae. aegypti* is widely present throughout the region, where warm temperatures remain suitable nearly year round, and summer monsoonal rains and human water storage near homes provides necessary breeding habitat. Studies from Tucson suggest that the highest densities of *Ae. aegypti* are found in neighborhoods with lower incomes and older homes (Walker *et al.*, 2011), as well in neighborhoods with medium height trees (Landau and van Leeuwen, 2012). Nevertheless, there have not been confirmed cases of dengue fever contracted within the state. However, there was an outbreak of travel-associated dengue in the southern city of Yuma, Arizona linked to an ongoing outbreak in nearby Sonora, Mexico (Jones *et al.*, 2016).

The situation across the border in neighboring Sonora, Mexico is different. Cases have been regularly reported in surveillance data within the state, with a notably large incidence reported in 2010 (see Ernst *et al.*, 2017). Regional variations in vegetation and access to piped water infrastructure may partly explain this trend (Hayden *et al.*, 2010; Jones *et al.*, 2016). However, it is also possible that climate may be limiting the lifespan of female *Ae. aegypti* in Arizona, reducing the potential of viral transmission to humans (Ernst *et al.*, 2017). There is also the concern that low-level transmission within Arizona may be going undetected, with the assumption that locally-acquired cases were contracted in Mexico, making increased physician awareness and surveillance important (Jones *et al.*, 2016; McCotter *et al.*, 2013).

The case of one Tucson resident helps to explicate the importance of physician awareness and surveillance in these regions. The resident, a middle-aged man without a travel history, reported to the Emergency Room with severe flu-like symptoms. He was ultimately discharged and told to take aspirin (contraindicated for dengue patients because of the risk of bleeding). He later presented to his primary care physician who tested for dengue. His physician interpreted his lab test as "positive" for dengue, with antibodies showing recent infection, but state health officials interpreted the results as inconclusive (Butterworth, personal communication, August 2012). This case occurred in 1999, and there have been no locally acquired dengue cases confirmed by state health officials in the state since. However, the possibility of undiagnosed dengue in the region remains an important question.

Changing ecologies and the future potential for dengue transmission in the US

Climate and socioeconomic conditions present in the southern US, and the gulf coast particularly (Hotez *et al.*, 2014), make it susceptible to a number of "tropical" diseases, including dengue. Poverty levels in each of the regions discussed are higher than the national average (*Table 1*). While this should not be interpreted as the sole determining factor of dengue, recognizing the risk this poses to disease transmission is crucial, and requires awareness and surveillance (Hotez, 2012). However, the capacities for surveillance and control are not uniform across the region, linked to larger state and federal structures that allocate funding to these activities (Hamer, 2016). As a whole, federal funding for arbovirus surveillance has decreased significantly over the past decade (CSTE, 2014) with vector control the second most negatively impacted activity, following food safety, as a result of such cuts (NACCHO, 2014). Florida's history of combating mosquitoes has developed into a comprehensive mosquito control program, which operates through mosquito control districts. However, the recent outbreaks of dengue fever, chikungunya, and most recently zika virus demonstrate that the state remains vulnerable to arboviral diseases. In the case of Arizona, mosquito-control activities are carried out by various agencies at the county level. A strain on resources in Pima County, for example, limits the control capacities of the local health department, while the comparatively better funded Maricopa County Vector Control district has a more active mosquito-control program (Shaw *et al.*, 2010). In looking at the heterogeneity of control resources across the United States, Hamer (2016, 494) notes that Texas is "vastly underfunded" in comparison to other states, despite a high risk for arbovirus transmission. All three states, through trade and travel, are connected to locations with endemic transmission in Mexico and/or the Caribbean.

Table 1. Median household income, per capita income, and percent of persons living in poverty in the US and selected counties referenced in this chapter in Florida, Texas, and Arizona (2011-2015).

	Median household income	Per capita income in past 12 months	Persons in poverty, percent
United States	$53,889	$28,930	13.5%
Pima County, AZ	$46,162	$25,729	18.7%
Yuma County, AZ	$40,743	$19,102	21.1%
Monroe County, FL	$57,290	$36,208	11.3%
Miami-Dade County, FL	$43,129	$23,850	20.0%
Cameron County, TX	$33,226	$15,105	32.0%
Harris County, TX	$54,457	$29,047	16.6%

Source: US census: www.census.gov/quickfacts/table/PST045216/00.

Attitudes towards the appropriate control measures for mosquitoes also vary. Rachel Carson's attention to the ecological harms of pesticides shifted the landscape of mosquito control within the US. This has drawn greater attention to the use of non-chemical options, including source reduction and biological controls. Nevertheless, in the epilogue of *The Mosquito Crusades* (which chronicles the development of the Mosquito Control movement within the US), Patterson (2009, 220) notes that, "despite these developments, mosquito control remains controversial. Public opposition to pesticides has intensified. Environmentalists continue to voice their concern about mosquito control's impact on biologically sensitive areas. At the same time, the control movement faces new challenges from mosquitoes and pathogens". These concerns are further apparent in the 2012 joint statement of the US Environmental Protection Agency and the Centers for Disease Control on the use of pesticides against mosquitoes. This document stresses the value of pesticides in protecting public health interests but recommends they be used alongside other non-chemical controls.

Such tradeoffs have been at the center of community debates, where residents question the larger ecosystem and health impacts of pesticide use. Over 8,000 signatures supported a petition (led by K. Litzenberger) in Dorchester County, South Carolina opposing mosquito pesticide spraying. In it, the petitioners cited concerns of pesticides on pollinators and general health and environmental impacts. Similar tensions have also been raised in the scholarly literature. The community of North Shore in Chicago, IL, slow to spray during the 2002 outbreak of West Nile virus, had faced previous opposition from residents regarding the use of pesticides to control mosquitoes owing to their environmental impacts (Tedesco *et al.*, 2010).

Such concerns have further been echoed in the three sites highlighted in this chapter. Studies in Tucson have found that residents there are generally reluctant about pesticide use owing environmental and human health concerns (Shaw *et al.*, 2010; von Hedemann *et al.*, 2017). In Texas, Chung *et al.* (2013, 306) noted that spraying during the 2012 Dallas WNV outbreak, "generated publicity over possible safety concerns," though human health impacts were not borne out in emergency room visits during the spray period. Another recent petition (led by D. Harris) against the spraying of Naled in south Florida in response to the Zika outbreak was signed by over 10,000 people concerned about the impact of the pesticide on human and pollinator health. In Key West, residents concerned about the environmental impacts of pesticides had a lower "willingness to pay" for mosquito control interventions (Dickinson *et al.*, 2016).

When considering human health exposures, the impacts of the pesticides on human health may be less than consequences of a vector borne-disease outbreak (Peterson *et al.*, 2006; Bonds, 2012). Nevertheless, as these examples highlight, the chemical control of mosquitoes versus its environmental and human health implications is fraught with tensions. The negative impacts of pesticide use have also spurred new alternatives, such as the production of sterile male mosquitoes, which also remain disputed within communities such as southern Florida (Palmer, 2015).

The tensions surrounding appropriate control measures may intensify in the southern US as climate change increases the potential for mosquito-borne disease outbreaks in the region. Simulations of mosquito activity (Monaghan *et al.*, 2016) suggest that *Ae. aegypti* has the potential to be active in the southernmost portions of Florida and Texas year round, and in Arizona from April through November. While the southern US represents the northern extension of dengue transmission in the Americas, climate change may facilitate conditions more conducive to viral transmission, both in its effects on mosquitoes and on the virus itself. Studies using global climate models (GCMs) have highlighted the potential for climate change to increase the intensity of dengue fever transmission in the US (e.g. Hales *et al.*, 2002). Within the southeastern US, it is projected that climate change may lengthen the seasonal period for active *Ae. aegypti* across the region, but the most notable increase in the possible transmission of dengue is in southern Florida, and along the Gulf Coast (Butterworth *et al.*, 2017). This is because while warmer temperatures may become amenable to *Ae. aegypti* survival across the region, the temperatures required by the virus remain more limited and are not likely to be uniformly suitable.

As with human adaptive capacities and climate, biological organisms can also change over time. *Ae. aegypti* is the predominant vector of dengue, but *Ae. albopictus* also spreads the disease, and was the vector implicated in the 2001-2001 dengue fever outbreak in Hawaii (Effler *et al.*, 2005). Both vectors are present in large portions of the southern US (CDC, 2017). In such places, interspecific competition can shift local patterns of vector abundance and ultimately impact local risk, given different vectorial capacities of the mosquitoes (see Morin *et al.*, 2013). The dengue virus has also evolved over time and space. As a result, some strains are now understood to cause more severe disease than others, such as the Southeast Asian strain of dengue virus type 2 (DEN-2)[4] and the Indian subcontinent strain of DEN-3 (Hesse, 2007). The evolution of more virulent strains, coupled with increased global movement poses

[4] Dengue infections are caused by four closely related viruses named DEN-1, DEN-2, DEN-3, and DEN-4.

greater chances for the introduction of these strains into new locations. More severe strains have displaced less severe strains in Central and South America, posing additional regional epidemiological concern (Rico-Hesse, 2003).

Teixeria *et al.* (2009, S8) remind us that, "It is important to understand the dynamics of the circulation of this virus in each geographic and social space." As ecosystems change and the dengue virus evolves, so too must our understandings of the 'geographic and social space' it occupies. The evidence presented in this chapter shows that dengue presents an ongoing concern throughout the southern US, and continued surveillance is important. At the same time, cases can and do go undetected, and resources needed to carry out important public health work may be less than adequate, and at times controversial. This is true even in comparatively well-resourced countries such as the US. At finer scales, risk may vary substantially, owing to differences in environmental influences, institutional capacities, and socioeconomic factors. It is unlikely that we will see dengue cases in the US that mirror historical outbreaks. Nevertheless, we must consider these complex socio-ecological relationships in order to understand the vulnerabilities to dengue fever within the southern region of the country.

Conclusion: Changing socio-environmental dynamics

The biologic, ecologic, and chemical control of mosquitoes over the past several centuries within the US has undoubtedly lessened the burden of mosquito-borne diseases. These changes have happened alongside improvements in built infrastructure that offer increased protective measures. At the same time, socio-ecological systems are not static. Those involved with environmental health are well attuned to the ways in which changes to environmental, biological, and behavioral conditions come together to lessen or amplify disease transmission. This chapter suggests that if we are to understand the possible re-emergence of dengue fever in the southern US, we must remember that the social and ecological conditions that govern dengue fever are continuously in flux. The control of malaria in the US was a great success, but the more recent epidemics of the West Nile virus and Zika viruses show that unique disease ecologies make some diseases more challenging to control than others. By fixing our assumptions of future risk based on parameters of the past, we miss this complexity and neglect opportunities to intervene. Climate change may expand the suitable habitat for *Ae. aegypti* and the transmission period for the dengue virus within the US. At the same time, the virus itself is capable of adapting and exploiting the movement of people and goods. All of these changes are set amidst the practical challenges of disease surveillance, diagnosis, and ultimately the provision of public health funding. Recognizing this complexity will help us more effectively combat deal with potential future outbreaks within the US, and increase our understanding of their spatial variation.

Acknowledgements

I'd like to extend many thanks to Lily House-Peters and Nicolena vonHedemann for their thoughtful comments and feedback on previous versions of this draft.

References

Adam J.K., Abeyta R., Smith B., Gaul L., Thomas D.L., Han G., … and Tomashek K.M. (2017). "Clinician Survey to Determine Knowledge of Dengue and Clinical Management Practices, Texas, 2014", *The American Journal of Tropical Medicine and Hygiene*, 96(3), 708-714.

Alley C. and Sommerfeld J. (2014). "Infectious disease in times of social and ecological change", *Medical anthropology*, 33(2), 85-91.

Anonymous (1997). The feasibility of eradicating *Aedes aegypti* in the Americas. *Pan Am J Public Health*, 1(1), PAHO 1997.

Beaumier C., Garcia M.N. and Murray K.O. (2014). "The history of dengue in the United States and its recent emergence", *Current Tropical Medicine Reports*, 1(1), 32-35.

Bhatt S., Gething P.W., Brady O.J., Messina J.P., Farlow A.W., Moyes C.L., … and Myers M.F. (2013). "The global distribution and burden of dengue", *Nature*, 496(7446), 504-507.

Bonds J.A.S. (2012). "Ultra-low-volume space sprays in mosquito control: a critical review", *Medical and Veterinary Entomology*, 26(2), 121-130.

Bouri N., Sell T.K., Franco C., Adalja A., Henderson D.A. and Hynes N.A. (2012). "Return of epidemic dengue in the United States: implications for the public health practitioner", *Public Health Reports*, 127(3),259-266.

Braga C., Luna C.F., Martelli C.M., De Souza W.V., Cordeiro M.T., Alexander N., … and Marques E.T. (2010). "Seroprevalence and risk factors for dengue infection in socio-economically distinct areas of Recife, Brazil", *Acta Tropica*, 113(3), 234-240.

Brunkard J.M. *et al.* (2007). "Dengue Fever Seroprevalence and Risk Factors, Texas–Mexico Border, 2004", *Emerging Infectious Disease Journal-CDC*, 13(10), 1477-1483.

Butterworth M. (2012, August). Email and personal interview (P. Howe, Tucson Resident).

Butterworth M. (2012, 2013). Personal Interviews (Key West Residents).

Butterworth M.K., Morin C.W. and Comrie A.C. (2017). "An analysis of the potential impact of climate change on Dengue transmission in the Southeastern United States", *Environ Health Perspect*, 125(4), 579-585.

Centers for Disease Control and Prevention (CDC) (2016a). "West Nile virus. Final Annual Maps and Data for 1999-2015". Online: https://www.cdc.gov/westnile/statsmaps/finalmapsdata/index.html. Accessed 4 March 2017.

Centers for Disease Control and Prevention (CDC) (2016b). "Malaria". Online: https://www.cdc.gov/malaria/about/history/index.html. Accessed 22 February 2017.

Centers for Disease Control and Prevention (CDC) (2012). "*Aedes aegypti* fact sheet". Online: https://www.cdc.gov/dengue/resources/30Jan2012/aegyptifactsheet.pdf. Accessed 6 March 2017.

Centers for Disease Control and Prevention (CDC) (2017). "Estimated range of *Aedes albopictus* and *Aedes aegypti* in the United States, 2016". Online: https://www.cdc.gov/zika/vector/range.html. Accessed 13 March 2017.

Chandra N. (2012). "Dengue devil on the prowl in South Delhi: control measures fail in posh areas that report 40% of Delhi cases, 22 October". Online: http://indiatoday.intoday.in/story/dengue-fever-zeroes-in-on-south-delhi-mcd-dengue-mosquito/1/225696.html. Accessed 1 September 2014.

Christophers S.R. (1960). *Aedes aegypti (L.), the yellow fever mosquito; its life history, bionomics, and structure.* Cambridge, Cambridge University Press.

Chung W.M., Buseman C.M., Joyner S.N., Hughes S.M., Fomby T.B., Luby J.P. and Haley R. (2013). The 2012 West Nile encephalitis epidemic in Dallas, Texas. *JAMA*, 310(3), 297-307.

Council of State and Territorial Epidemiologists (CSTE) (2014). "Assessment of Capacity in 2012 for the Surveillance, Prevention and Control of West Nile Virus and other Mosquito-borne Virus Infections in State and Large City/County Health Departments and How it Compares to 2004". Online: http://www.cste2.org/docs/VBR.pdf. Accessed: 4 March 2017.

Craddock S. and Brown T. (2009). *Representing the un/healthy body* (pp. 301-321). Oxford: Wiley-Blackwell.

Darsie J. and Ward R. (2005). *Identification and geographical distribution of the mosquitoes of North America and North of Mexico*. University of Florida.

Dickinson K.L., Hayden M.H., Haenchen S., Monaghan A.J., Walker K.R. and Ernst K.C. (2016). "Willingness to Pay for Mosquito Control in Key West, Florida and Tucson, Arizona", *The American Journal of Tropical Medicine and Hygiene*, 94(4), 775-779.

Effler P., Pang L., Kitsutani P., Vorndam V., Nakata M., Ayers M. *et al.* (2005). "Dengue Fever, Hawaii, 2001-2002", *Emerg Infect Dis.*, 11(5), 742-749.

Eisen L. and Moore C.G. (2013). "Aedes (Stegomyia) aegypti in the continental United States: a vector at the cool margin of its geographic range", *Journal of Medical Entomology*, 50(3), 467-478.

Ehrenkranz N.J., Ventura A.K., Cuadrado R.R., Pond W.L. and Porter J.E. (1971). Pandemic dengue in Caribbean countries and the southern United States–past, present and potential problems. *New England Journal of Medicine*, *285*(26), 1460-1469.

Environmental Protection Agency and Centers for Disease Control (2012). "Joint statement on Mosquito Control in the United States". Online: https://www.epa.gov/mosquitocontrol/joint-statement-mosquito-control-united-states. Accessed 4 May 2017.

Ernst K.C., Haenchen S., Dickinson K., Doyle M.S., Walker K., Monaghan A.J. and Hayden M.H. (2015). "Awareness and support of release of genetically modified 'sterile' mosquitoes, key west, Florida, USA", *Emerging Infectious Diseases*, 21(2), 320.

Ernst K.C., Walker K.R., Reyes-Castro P., Joy T.K., Castro-Luque A.L., Diaz-Caravantes R.E., … and Jeffrey-Guttierez E. (2017). "Aedes aegypti (Diptera: Culicidae) Longevity and Differential Emergence of Dengue Fever in Two Cities in Sonora, Mexico", *Journal of Medical Entomology*, 54(1), 204-211.

Fazlulhaq N. (2012). "Rich or Poor, Dengue Kills", 10 June. Online: http://www.sundaytimes.lk/120610/News/nws_44.html. Accessed 1 September 2014.

Florida Department of Health (2017). "Dengue". Online: http://www.floridahealth.gov/diseases-and-conditions/dengue/. Accessed 16 March 2017.

Gale J. and Srivastava M. (2010). "In India, dengue fever stalks the affluent". 16 September. Online: https://www.bloomberg.com/news/articles/2010-09-16/in-india-dengue-fever-stalks-the-affluent. Accessed 13 March 2017.

Gubler D.J. and Clark G.G. (1995). "Dengue/dengue hemorrhagic fever: the emergence of a global health problem", *Emerging infectious diseases*, 1(2), 55.

Guha-Sapir D. and Schimmer B. (2005). "Dengue fever: new paradigms for a changing epidemiology", *Emerging Themes in Epidemiology*, 2(1), 1.

Hales S., De Wet N., Maindonald J. and Woodward A. (2002). "Potential effect of population and climate changes on global distribution of dengue fever: an empirical model", *The Lancet*, *360*(9336), 830-834.

Hamer G.L. (2016). "Heterogeneity of Mosquito (Diptera: Culicidae) Control Community Size, Research Productivity, and Arboviral Diseases Across the United States", *Journal of Medical Entomology*, 53(3), 485-495.

Harris D. (n.d.). "Petition: Floridians against spraying of Naled". Online: http://www.thepetitionsite.com/630/738/913/floridians-against-spraying-of-naled/?taf_id=28774804&cid=fb_na#bbfb=745879123. Accessed 5 May 2017.

Hayden M.H., Uejio C.K., Walker K., Ramberg F., Moreno R., Rosales C., … and Janes C.R. (2010). "Microclimate and human factors in the divergent ecology of Aedes aegypti along the Arizona, US/Sonora, MX border", *EcoHealth*, 7(1), 64-77.

Hayden M.H., Cavanaugh J.L., Tittel C., Butterworth M., Haenchen S., Dickinson K., … and Ernst K.C. (2015). "Post outbreak review: Dengue preparedness and response in Key West, Florida", *The American Journal of Tropical Medicine and Hygiene*, 93(2), 397-400.

Hayes Jr G.R., Scheppf P.P. and Johnson E.B. (1971). "An historical review of the last continental US epidemic of dengue", *Mosquito News*, 31(3), 422-427.

Hesse R.R. (2007). "Dengue virus evolution and virulence models", *Clinical Infectious Diseases*, 44(11), 1462-1466.

Hotez P.J., Murray K.O. and Buekens P. (2014). "The Gulf Coast: a new American underbelly of tropical diseases and poverty", *PLoS Negl Trop Dis*, 8(5), e2760.

Hotez P.J. (2012). "Fighting neglected tropical diseases in the southern United States", *BMJ-British Medical Journal*, 345(7875), 10.

Hotez P.J. (2016). "Zika in the United States of America and a Fateful 1969 Decision", *PLoS Negl Trop Dis*, 10(5), e0004765.

Jones J.M., Lopez B., Adams L., Galvez F., Nunez A., Santillan N., … and Komatsu K. (2016). Binational Dengue Outbreak Along the United States–Mexico Border–Yuma County, Arizona, and Sonora, Mexico, 2014. *MMWR. Morbidity and Mortality Weekly Report, 65*.

Kuno G. (2015). "A re-examination of the history of etiologic confusion between dengue and chikungunya", *PLoS Negl Trop Dis*, 9(11), e0004101.

Landau K.I. and van Leeuwen W.J. (2012). "Fine scale spatial urban land cover factors associated with adult mosquito abundance and risk in Tucson, Arizona", *Journal of Vector Ecology*, 37(2), 407-418.

Litzenberger K. (n.d.). "Petition: Save Humanity Save the Bees". Online: https://www.change.org/p/dorchester-county-and-the-town-of-summerville-save-humanity-save-the-bees?recruiter=590397671&utm_source=share_for_starters&utm_medium=copyLin. Accessed 4 May 2017.

McCotter O., Vanskike F., Ernst K., Komatsu K., Margolis H., Waterman S., … and Hunsperger E. (2013). "Enhancing Surveillance for Arboviral Infections in the Arizona Border Region". *Online Journal of Public Health Informatics*, 5(1), e4.

Monaghan A.J., Morin C.W., Steinhoff D.F., Wilhelmi O., Hayden M., Quattrochi D.A., … and Scalf P.E. (2016). "On the seasonal occurrence and abundance of the Zika virus vector mosquito Aedes aegypti in the contiguous United States", *PLoS Currents*, 8.

Morin C.W., Comrie A.C. and Ernst K. (2013). "Climate and dengue transmission: evidence and implications", *Environmental Health Perspectives (Online)*, 121(11-12), 1264.

Mulligan K., Elliott S.J. and Schuster-Wallace C.J. (2012a). "Global public health policy transfer and dengue fever in Putrajaya, Malaysia: A critical discourse analysis", *Critical Public Health*, 22(4), 407-418.

Mulligan K., Elliott S.J. and Schuster-Wallace C. (2012b). "The place of health and the health of place: dengue fever and urban governance in Putrajaya, Malaysia", *Health and Place*, 18(3), 613-620.

Mulligan K., Dixon J., Joanna Sinn C.L. and Elliott S.J. (2015). "Is dengue a disease of poverty? A systematic review", *Pathogens and Global Health*, 109(1), 10-18.

Murray K.O., Rodriguez L.F., Herrington E., Kharat V., Vasilakis N., Walker C., ... and Martinez D. (2013). "Identification of dengue fever cases in Houston, Texas, with evidence of autochthonous transmission between 2003 and 2005", *Vector-Borne and Zoonotic Diseases*, 13(12), 835-845.

National Association of County and City Health Officials (NACCHO) (2014). "Impact of Budget Cuts on Environmental Health Services at Local Health Departments". Online: http://essentialelements.naccho.org/wp-content/uploads/2016/05/Impact-of-Budget-Cuts-on-Environmental-Health-Services-at-LHDs-1.pdf. Accessed 4 March 2017.

Omran A. (1971). "The epidemiologic transition: A theory of the epidemiology of population change", *The Millbank Quarterly*, 83(4), 731-757.

Palmer L. 2015. "Genetically Modified Mosquito Sparks a Controversy in Florida". Online: http://e360.yale.edu/features/genetically_modified_mosquito_sparks_a_controversy_in_florida. Accessed 4 May 2017.

Pan American Health Organization (PAHO) (2013). "Epidemiologic Alert: Dengue 21 June 2013". Online: http://www.paho.org/hq/index.php?option=com_docman&task=doc_view&gid=22101&Itemid=270&lang=en. Accessed 16 March 2017.

Pan American Health Organization (PAHO) (2017). "2016: Numbers of dengue and severe dengue (SD) in the Americas by Country – January 26, 2017 (EW 52)". Online: http://www.paho.org/hq/index.php?option=com_topics&view=rdmore&cid=6290&Itemid=40734 Accessed 16 March 2017.

Patterson G. (2004). *The mosquito wars: a history of control in Florida*. University Press of Florida.

Patterson G. (2009). *The mosquito crusades: A history of the American anti-mosquito movement from the reed commission to the first earth day*. Rutgers University Press.

Patterson K.D. (1992). "Yellow fever epidemics and mortality in the United States, 1693-1905", *Social Science and Medicine*, 34(8), 855-865.

PETERSON, R. K., MACEDO, P. A., and DAVIS R. S. (2006). "A human-health risk assessment for West Nile virus and insecticides used in mosquito management", *Environmental Health Perspectives*, 366-372.

RADKE E.G. (2012). "Dengue Outbreak in Key West, Florida, USA, 2009", *Emerging Infectious Disease Journal-CDC*, 18(1).

RAMOS M.M., MOHAMMED H., ZIELINSKI-GUTIERREZ E., HAYDEN M.H., LOPEZ J.L.R., FOURNIER M., … and BANICKI A.A. (2008). "Epidemic dengue and dengue hemorrhagic fever at the Texas–Mexico border: results of a household-based seroepidemiologic survey, December 2005", *The American Journal of Tropical Medicine and Hygiene*, 78(3), 364-369.

REITER P., LATHROP S., BUNNING M.L., BIGGERSTAFF B.J., SINGER D., TIWARI T., … and HAYES E. (2003). "Texas Lifestyle Limits Transmission of Dengue Virus", *Emerging Infectious Diseases*, 9(1), 86-89. Online: https://dx.doi.org/10.3201/eid0901.020220.

RICO-HESSE R. (2003). "Microevolution and virulence of dengue viruses", *Advances in Virus Research*, 59, 315-341.

SHAW I.G.R., ROBBINS P.F. and JONES III J.P. (2010). "A bug's life and the spatial ontologies of mosquito management", *Annals of the Association of American Geographers*, 100(2), 373-392.

SOPER F.L. (1963). "The elimination of urban yellow fever in the Americas through the eradication of *Aedes aegypti*", *American Journal of Public Health and the Nations Health*, 53(1), 7-16.

SOPER F.L. (1967). "Dynamics of Aedes aegypti distribution and density. Seasonal fluctuations in the Americas", *Bulletin of the World Health Organization*, 36(4), 536.

TEDESCO C., RUIZ M. and MCLAFFERTY S. (2010). "Mosquito politics: local vector control policies and the spread of West Nile Virus in the Chicago region", *Health and Place*, 16(6), 1188-1195.

TEIXEIRA M.D.G., BARRETO M.L., COSTA M.D.C.N., FERREIRA L.D.A., VASCONCELOS P.F. and CAIRNCROSS S. (2002). "Dynamics of dengue virus circulation: a silent epidemic in a complex urban area", *Tropical Medicine and International Health*, 7(9), 757-762.

TEIXEIRA M.G., COSTA M.D.C.N., BARRETO F. and BARRETO M.L. (2009). "Dengue: twenty-five years since reemergence in Brazil", *Cadernos de saúde pública*, 25, S7-S18.

Thomas D.L., Santiago G.A., Abeyta R., Hinojosa S., Torres-Velasquez B., Adam J.K., ... and Smith B. (2016). "Reemergence of dengue in Southern Texas, 2013", *Emerging Infectious Diseases*, 22(6), 1002.

United States Geological Survey (USGS) (2016). "Disease Maps Dengue Virus (Locally Acquired)". Online: https://diseasemaps.usgs.gov/mapviewer. Accessed 16 March 2017.

von Hedemann N., Robbins P., Butterworth M.K., Landau K. and Morin C.W. (2017). "Managing mosquito spaces: Citizen self-governance of disease vectors in a desert landscape", *Health and Place*, 43, 41-48.

Walker K.R., Joy T.K., Ellers-Kirk C. and Ramberg F.B. (2011). "Human and environmental factors affecting *Aedes aegypti* distribution in an arid urban environment", *Journal of the American Mosquito Control Association*, 27(2), 135-141.

World Health Organization (2017). "Dengue Control, Epidemiology". Online: http://www.who.int/denguecontrol/epidemiology/en/. Accessed 17 March 2017.

Climate change and Malaria in Burkina Faso

Eric Diboulo

Senior Epidemiologist/Biostatistician, Nouna Health Reserch Center (CRSN), Burkina Faso

It is now a proven fact that greenhouse gas emissions from human activity are a major contributor to global climate change. The return values of extreme weather and climate events are among the most consequential changes anticipated, since they are indeed expected to become substantially shorter over the course of this century (IPCC, 2012).

Climate change coupled with changing demographics is expected to magnify the already significant adverse effects of extreme weather on public health (McMichael *et al.*, 2006). On average, the Earth's climate has warmed by about 0.6°C over the past 100 years, with temperature increases especially pronounced since the mid-1970s (IPCC, 2012). Furthermore the vulnerability of populations to heat-related mortality is often characterized and modified by the underlying prevalence of temperature-sensitive diseases, socioeconomic development, and the age structure of the population (McMickael *et al.*, 2006). Research has reported short-term connections between rainfall and mortality (WHO, 2016). Tropical, vector-borne diseases, such as the biting rate of malaria mosquitoes and related human incidence rates, are exacerbated shortly after a rainfall event (Yé *et al.*, 2009). Based on average climate outcomes for the period from 1961-1990, World Health Organization (WHO) currently estimates that > 150,000 deaths and a burden of 5.5 million DALYs can be attributed to climate change and climate variability each year (WHO, 2014).

While predicted risk profiles for climate-sensitive diseases tend to broadly forecast a worsening in most parts of the world, it is suspected that low income and emerging economy countries will be more severely affected than high-income countries (IPCC, 2012). For example, WHO estimates that the environment-related disease burden will be much greater in low income countries than in high income countries (25% versus 17% of deaths) (IPCC, 2012). WHO further estimates that warming and precipitation trends due to anthropogenic climate change over the past 30 years already claim over 150,000 lives annually (WHO, 2014).

Weather variables and the biological cycle of malaria transmission

Malaria infection occurs by exposure to the bites of infective female mosquitoes of the *Anopheles* genus. The biological cycle of malaria transmission involves the alternation between human host and mosquito vector. Of the four malaria parasite species, *Plasmodium falciparum* is the most common to affect humans and is found throughout most sub-Saharan African counties (Hay *et al.*, 2005). Environmental and climatic conditions – particularly temperature and rainfall – are the main drivers of malaria transmission, directly affecting the ability of the parasites and anopheline vector species to coexist long enough to enable and sustain transmission. Indeed, successful development of the malaria parasite in mosquitoes depends on several factors, the most important of which are temperature and humidity (higher temperatures accelerate the parasite's growth in the mosquitoes), and whether *Anopheles* survives long enough to allow the parasite to complete its cycle in the mosquito host.

The development of the parasite within the mosquito (sporogonic cycle) is greatly influenced by temperature. The sporogonic cycle takes about 9 to 10 days at temperatures of 28°C, but stalls at temperatures below 16°C (Yé *et al.*, 2007). Temperature is a key determinant in the daily survival of the vector. At temperatures between 16°C and 36°C, the daily survival rate is about 90%. This rate drops rapidly at temperatures above 36°C. The highest proportion of vectors surviving the incubation period is observed at temperatures between 28° and 32°C (Craig, Snow and Le Sueur, 1999). The gonotrophic cycle – i.e., the time between two blood meals for the vector – is significantly shortened at higher temperatures by speeding up the digestion process (Detinova, Bertram and Organization, 1962). Therefore, higher temperatures result in more frequent vector-host contact.

Consequently, by provoking extreme weather conditions, climate change is expected to have a marked impact on weather-related morbidity and mortality ("Impact of Regional Climate Change on Human Health", 2017). At present, however, knowledge and quantitative estimates of the impact of weather and extreme climatic events on vector-borne diseases, especially malaria, are still sparse – particularly for sub-Saharan Africa.

Climate change and malaria transmission in Burkina Faso

Country profile

Burkina Faso lies mostly between latitudes 9° and 15°N and longitudes 6°W and 3°E (*Figure 1*). It is made up of two major types of countryside: the larger part of the country is covered by a "peneplain", which forms gently undulating landscapes with a few isolated hills in some areas; the southwest of the country forms a sandstone massif bordered with sheer cliffs up to 150 meters high. Burkina Faso is therefore a relatively flat country with an average altitude of around 400 meters. Four main rivers drain the country: the Mouhoun, the Nakambé, the Nazinonand and the Komoé. The Mouhoun is one of only two rivers in the country that flows year-round, the other being Komoé, which flows to the southwest. Burkina Faso has a primarily tropical climate with two very distinct seasons. In the rainy season, the country receives between 600 and 900 millimetres of rainfall; malaria is known for a seasonal recrudescence in this period during which it is the main cause of fever and mortality in the country. The rainy season lasts approximately four months, from May/June to September, and is shorter in the north of the country. In the dry season, the "harmattan" – a hot dry wind – blows from the Sahara, carrying dust and dirt that contribute to high morbidity from lower respiratory infections. As of January 2017, the population of Burkina Faso was estimated at 18,909,665, which represents an increase of 2.93% compared to the 2016 population.

The economy of Burkina Faso is heavily reliant on agriculture, which employs close to 80% of the active population. Cotton is the country's most important cash crop, while gold exports have gained importance in recent years. The opening of new industrial mines, plus a slight rebound in gold and cotton prices, and relatively good agricultural production after relatively good rainfall all contributed to a gradual return to economic growth in 2016. The gross domestic product (GDP) grew 5.4%, well above the rate of 4% observed in 2014 and 2015 (Worldbank, 2017).

Figure 1. *Burkina Faso*

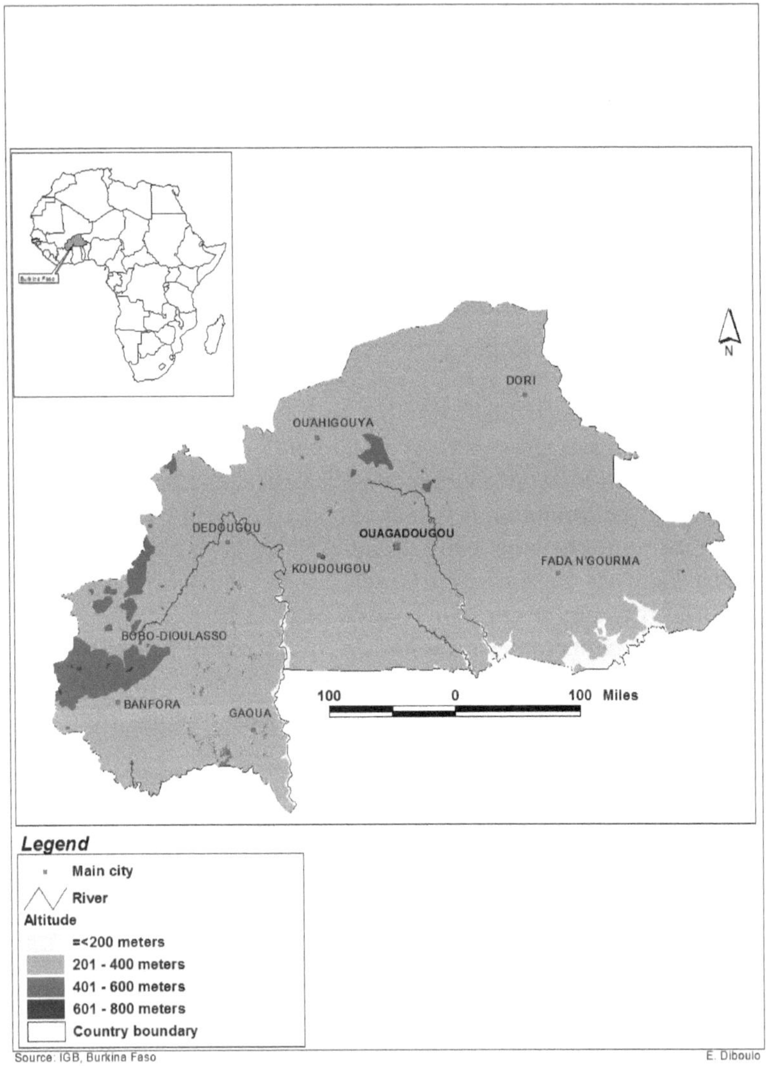

Malaria morbidity and mortality in Burkina Faso

Malaria is holoendemic in Burkina Faso with most transmission occurring during or shortly after the rainy season between July and December. Ninety-nine percent of infection is attributed to *P. falciparum*. The overall prevalence of infection in children aged 6-59 months is estimated at 66% (EDSBF-MICS IV, 2010). The government made tremendous efforts to meet the objectives of the 2006-2010 National Malaria Strategic Plan and implemented special programs such as: home-based malaria management in 2008; universal coverage of insecticide-treated bed nets (ITN) in 2010; intermittent preventive therapy (IPT) for high-risk groups in 2005; piloting indoor residual spray (IRS) in certain health districts since 2010; larval control and sanitation programs; the introduction of effective tools for malaria control, notably rapid diagnostic tests (RDTs) at all health facilities in 2010; and the actual availability of artemisinin-based combination therapies (ACTs) in health facilities in 2007 (Burkina Faso Ministry of Health, 2015).

The percentage of deaths attributable to malaria in children under the age of 5 decreased from 8.1% in 2000 to 3.3% in 2010 (Burkina Faso Ministry of Health, 2015). This decline can partly be attributed to the success of a wide range of malaria interventions and control programs such as insecticide treated nets (ITN); the actual availability of ACTs in health facilities (2007); the launch of a home-based malaria management (Burkina Faso Ministry of Health, 2008); the introduction of RDTs as part of malaria case management at all health facilities (2010); universal coverage of ITNs (2010) and IPT for high-risk groups as part of the 2006-2010 malaria strategic plan. However, data from the 2010 Burkina Faso Health and Demographic Survey-Malaria Indicator Cluster Survey (DHS-MICS) used in Bayesian geostatistical modeling of infant and under-five mortality in relation to malaria endemicity concluded that, in terms of levels of differentials in infant and child mortality based on certain socioeconomic and demographic factors and place of residence, there is a significant reduction in the mortality hazard in children born in wealthy households. While this hazard decreases with higher socio-economic status in rural areas, only children born in the richest households (fifth SES quintile) have a significantly reduced risk of mortality in urban areas (Diboulo *et al.*, under review). Infants born to older mothers (between 20-29 and 30-39 years old) in rural settings are at significantly reduced mortality hazard. However, being born into a large household in a rural setting significantly increases the infant mortality hazard ratio (HR) (HR=1.37; 95% CI: 1.00-1.88). Infants born in households with 3 or more children under five in urban areas have a significantly lower mortality rate (HR=0.23; 95% CI: 0.10-0.48); this also applies to both infants and children in rural settings who respectively face a HR of (0.47; 95% CI: 0.37-0.59) and (0.49; 95% CI: 0.37-0.66). Infants and children born with a large birth interval are also less likely to die in rural areas. There is an important negative correlation between birth size and risk of infant death in both rural and urban areas (HR=0.39; 95% CI: 0.36-0.93 and HR=0.68; 95% CI: 0.53-0.89, respectively). This association is also important for children living in rural areas (HR=0.66; 95% CI: 0.46-0.95).

Furthermore, in rural settings, the risk of infant mortality increases by 12% for each increase of birth order by 1 child (HR=1.12; 95% CI: 1.06-1.18). The mortality rate is significantly lower among infants born to mothers who had at least one antenatal visit during pregnancy (HR= 0.77; 95% CI: 0.05-0.11 for rural areas and HR=0.12; 95% CI: 0.10-0.15 for urban areas) and for children in rural settings, HR= 0.69 (95% CI: 0.54-0.88). Infants and children born from multiple births have a significantly higher mortality rate in rural settings HR = 2.39 (95% CI: 1.87-3.04) and 2.23 (95% CI: 1.34-3.54). The mortality rate is also significantly higher among children born in urban settings where malaria prevalence is above 75%, HR = 3.20 (95% CI: 1.01-10.46).

Rainfall and malaria transmission in Burkina Faso

A number of studies have looked at the effect of weather and climate variables on malaria transmission from both an entomological and parasitological perspective through different modeling approaches (Samadoulougou *et al.*, 2014; Diboulo *et al.*, 2015; Diboulo *et al.*, 2016), while others have focused on detecting the correlation between weather conditions and malaria morbidity and mortality in Burkina Faso (Diboulo *et al.*, 2012).

There have been many attempts to quantify how variability in climate factors affects malaria transmission. A number of studies have reported a significant correlation between increased precipitation and malaria transmission (Nebie *et al.*, 2008; Bisoffi *et al.*, 2010; Diarra *et al.*,

2012). Nationally representative survey data used in multi-level geo-statistical modeling has however shown a negative correlation between rainfall and malaria parasitaemia (Samadoulougou *et al.*, 2014).

Research by Diboulo *et al.* (2015) using Bayesian variable selection to model geographical heterogeneity in malaria transmission using sparse data from the Nouna Health and Demographic Surveillance System (HDSS) in Burkina Faso found a correlation between rainfall and the densities of both *An. gambiae* and *An. funestus*. The direction of the effect was different for the different species, however; indeed, the density of *An. gambiae* was found to be negatively correlated with rainfall while a positive correlation was found with the density of *An. funestus*. The negative correlation with *An. gambiae* density may suggest that although rainfall remains an important factor for the development of this species, consecutive heavy rainfall (over the current and two previous months) may flush away all suitable *An. gambiae* breeding sites; thus, for *An. gambiae*, the most efficient malaria vector is a rainy-dependant species which favors temporary and shallow breeding sites. The positive correlation with *An. funestus* density indicates that rainfall is important for the development and survival of this species, which predominantly develops in permanent water bodies with emerging vegetation (Dia, Wamdaogo and Ayal, 2013).

Cumulative rainfall from the previous month was found to have a positive effect on clinical malaria rates in Burkina Faso (Yé *et al.*, 2007). This finding is consistent with that of several previous studies (Teklehaimanot *et al.*, 2004; Zhou *et al.*, 2005; Githeko and Ndegwa, 2001; Minakawa *et al.*, 2002). The relationship was further found to be "J-shaped".[5] The J-shape is characteristic of the dose-response curve. J-shaped curves usually describe a type of effect in which after a certain "threshold", even a small increase in the dose results in a marked response. Furthermore, the study found that the minimum amount of rainfall necessary to trigger the transmission was 100 mm. This amount was necessary before any case of clinical malaria could be observed. Potential explanations for this finding involve the hot and dry climate of the study area, with high evaporation and water infiltration into the soil. Small amounts of rainfall evaporate or infiltrate quickly; conversely, with greater amounts of rainfall, some water pools for an extended period,

5 A J curve is any variety of J-shaped diagram where a curve initially falls, then steeply rises above the starting point.

thus providing suitable breeding grounds for the vector to complete its development cycle.

Temperature and malaria transmission in Burkina Faso

Temperature has been associated with the dynamics of malaria vector population and, therefore, with transmission of the disease. Ambient temperature plays a major role in the life cycle of the malaria vector. A study that aimed to assess the effects of meteorological factors on clinical risk among children in Burkina Faso found that mean temperature was a strong predictor of clinical malaria rates in children (Yé *et al.*, 2007). Moreover, the bell-shaped relationship between temperature and malaria infection risk in that study suggests lower risks at low temperatures and higher risks with increasing temperatures. The infection risk was found to be highest at temperatures around 27°C. A recent study in Burkina Faso using nationally representative household malaria survey data from 2010-2011 (Diboulo *et al.*, 2016) found a negative correlation between increased night temperature and malaria transmission. This is consistent with laboratory experiments that have observed the shortest *Anopheles gambiae s.s* larval survival (<7 days) at 10–12 and 38–40 °C and the highest larval mortality occurring between 30 and 32°C, with death (rather than adult emergence) representing over 70% of the terminal events in mosquitoes originally from Lagos (Nigeria) (Bayoh and Lindsay, 2004). In Burkina Faso, the monthly mean temperature in the hottest and driest period (March-May) is constantly well above 31°C. Land surface temperature at night therefore appears to be an important predictor of malaria transmission. Furthermore, the behavioral high temperature avoidance experiment showed that *An. gambiae*, the most efficient malaria vector species in Burkina Faso, was more sensitive to increased temperatures than its sibling species, *Anopheles arabiensis* (Bayoh and Lindsay, 2004). A significant negative correlation between high temperatures [38.5-39.5°C] and malaria infection was also found in a previous modeling effort in Burkina Faso (Samadoulougou *et al.*, 2014).

Another recent modeling effort in Burkina Faso (Diboulo *et al.*, 2015) found a negative correlation between night temperatures and *An. gambiae* mosquito density more specifically. These findings possibly imply that although the high temperatures of the study area (average daily minimum: 20-28.1°C, maximum: 29.5-37.2°C) are suitable for stable malaria transmission (Craig, Snow and Le Sueur, 1999), relief from the heat mainly at night is also a key determinant in mosquito development and survival.

Most future climate change scenarios ultimately predict extreme weather conditions such as increased temperature and rainfall. We can therefore expect to see a marked impact on vector-borne diseases such as malaria, as both increases in temperature and rainfall are likely to provide more conducive environments for mosquitoes to thrive and extend malaria transmission to areas that have thus far been spared.

The recent growth of spatial data availability and climatic models at any scale and worldwide have spurred a renewed interest in modeling malaria health outcomes. This growth and availability provides ground for the first model intercomparison for health impacts in a future affected by climate change.

A recent study (Caminade *et al.*, 2014) used bias-corrected temperature and rainfall simulations from the Coupled Model Intercomparison Project Phase 5 climate models to compare the metrics of five statistical and dynamical malaria impact models for three future time periods. This study further evaluated three malaria outcome metrics at the global and regional levels. The findings showed an overall global net increase in the population at risk. Climate change resulting in extreme weather conditions (which is the most likely scenario according to most climate models) is therefore expected to exacerbate malaria transmission in most parts of sub-Saharan Africa, and particularly in Burkina Faso.

Potential impacts from climate change involve reversing the current downward trend observed over the last decade at the global level (WHO, 2016) in terms of malaria morbidity and mortality, thus compromising achievement of the sustainable development goals (SDGs). Malaria-endemic countries should take account of this and further adapt country-specific national climate change adaption/ mitigation plans in order to build more resilient health systems.

Conclusion

Emerging evidence (climate models and projections) suggests that future climate change is likely to result in extreme weather conditions (heat waves and heavy rainfall). Furthermore, a growing body of evidence consistently points to the likely impacts of extreme weather conditions on malaria transmission in sub-Saharan Africa and particularly in Burkina Faso. The effects of future climate change are therefore expected to exacerbate malaria transmission in Burkina Faso. Future research and modeling efforts should focus on predicting the effectiveness of the current malaria interventions under future climate change scenarios in order to ascertain the level of additional effort required to maintain current levels of malaria control or to achieve a more significant reduction in morbidity, as well as the number of deaths that could be prevented.

References

BAYOH M.N. and LINDSAY S.W. (2004). "Temperature-Related Duration of Aquatic Stages of the Afrotropical Malaria Vector Mosquito Anopheles Gambiae in the Laboratory", *Medical and Veterinary Entomology*, 18(2), 174-179. https://doi.org/10.1111/j.0269-283X.2004.00495.x.

BISOFFI Z., SIRIMA S.B., MENTEN J., PATTARO C., ANGHEBEN A., GOBBI F., TINTO H. *et al.* (2010). "Accuracy of a Rapid Diagnostic Test on the Diagnosis of Malaria Infection and of Malaria-Attributable Fever during Low and High Transmission Season in Burkina Faso", *Malaria Journal*, 9(July), 192. https://doi.org/10.1186/1475-2875-9-192.

Burkina Faso Ministry of Health, *Plan Stratégique 2011_2015 Palu* (n.d.). Online: http://www.sante.gov.bf/index.php/quotidien/actualites/210-plan-strategique-20112015-palu. Accessed July 20, 2015.

CAMINADE C., KOVATS S., ROCKLOV J., TOMPKINS A.M., MORSE A.P., COLÓN-GONZÁLEZ F.J., STENLUND H., MARTENS P. and LLOYD S.J. (2014). "Impact of Climate Change on Global Malaria Distribution", *Proceedings of the National Academy of Sciences of the United States of America*, 111(9), 3286-3291. https://doi.org/10.1073/pnas.1302089111.

CDC-Centers for Disease Control and Prevention "CDC – Malaria – About Malaria – Biology." (n.d.). Online: https://www.cdc.gov/Malaria/about/biology/index.html. Accessed July 18, 2017.

"Climate Change and Human Health: Present and Future Risks." – PubMed – NCBI (n.d.). Online: https://www.ncbi.nlm.nih.gov/pubmed/16530580. Accessed July 21, 2017.

CRAIG M.H., SNOW R.W. and LE SUEUR D. (1999). "A Climate-Based Distribution Model of Malaria Transmission in Sub-Saharan Africa", *Parasitology Today*, 15(3), 105-111. https://doi.org/10.1016/S0169-4758(99)01396-4.

Detinova T.S., Bertram D.S. and World Health Organization (1962). "Age-Grouping Methods in Diptera of Medical Importance, with Special Reference to Some Vectors of Malaria". Online: http://www.who.int/iris/handle/10665/41724.

Dia I., Wamdaogo M. and Ayal D. (2013). "Advances and Perspectives in the Study of the Malaria Mosquito Anopheles Funestus", in *Anopheles Mosquitoes – New Insights into Malaria Vectors*, edited by Sylvie Manguin. Rijeka, Croatia: InTech. Online: http://www.intechopen.com/books/references/anopheles-mosquitoes-new-insights-into-malaria-vectors/advances-and-perspectives-in-the-study-of-the-malaria-mosquito-anopheles-funestus.

Diarra A., Nébié I., Tiono A., Sanon S., Soulama I., Ouédraogo A., Gansané A. *et al.* (2012). "Seasonal Performance of a Malaria Rapid Diagnosis Test at Community Health Clinics in a Malaria-Hyperendemic Region of Burkina Faso", *Parasites and Vectors*, 5(May), 103. Online: https://doi.org/10.1186/1756-3305-5-103.

Diboulo E., Sié A., Diadier D.A., Karagiannis Voules D.A., Yé Y. and Vounatsou P. (2015). "Bayesian Variable Selection in Modelling Geographical Heterogeneity in Malaria Transmission from Sparse Data: An Application to Nouna Health and Demographic Surveillance System (HDSS) Data, Burkina Faso", *Parasites and Vectors*, 8(February), 118. Online: https://doi.org/10.1186/s13071-015-0679-7.

Diboulo E., Sié A., Rocklöv J., Niamba L., Yé M., Bagagnan C. and Sauerborn R. (2012). "Weather and Mortality: A 10 Year Retrospective Analysis of the Nouna Health and Demographic Surveillance System, Burkina Faso", *Global Health Action*, 5(0). Online: https://doi.org/10.3402/gha.v5i0.19078.

Diboulo E., Sié A. and Vounatsou P. (2016). "Assessing the Effects of Malaria Interventions on the Geographical Distribution of Parasitaemia Risk in Burkina Faso", *Malaria Journal*, 15(April), 228. Online: https://doi.org/10.1186/s12936-016-1282-x.

Githeko A.K. and Ndegwa W. (2001). "Predicting Malaria Epidemics in the Kenyan Highlands Using Climate Data: A Tool for Decision Makers", *Global Change and Human Health*, 2(1), 54-63. Online: https://doi.org/10.1023/A:1011943131643.

Hay S.I., Guerra C.A., Tatem A.J., Atkinson P.M. and Snow R.W. (2005). "Urbanization, Malaria Transmission and Disease Burden in

Africa", *Nature Reviews. Microbiology*, 3(1), 81-90. Online: https://doi.org/10.1038/nrmicro1069.

Institut National de la Statistique et de la Démographie (Burkina Faso and ICF International USA) (2012). *Enquête démographique et de santé et à indicateurs multiples du Burkina Faso 2010 (EDSBF-MICS IV)*. Ouagadougou (BFA); Calverton, Maryland (USA), INSD, ICF International.

Minakawa N., Sonye G., Mogi M., Githeko A. and Yan G. (2002). "The Effects of Climatic Factors on the Distribution and Abundance of Malaria Vectors in Kenya", *Journal of Medical Entomology*, 39(6), 833-841.

Nebie I., Tiono A.B., Diallo D.A., Samandoulougou S., Diarra A., Konate A.T., Cuzin-Ouattara N. *et al.* (2008). "Do Antibody Responses to Malaria Vaccine Candidates Influenced by the Level of Malaria Transmission Protect from Malaria?", *Tropical Medicine and International Health: TM and IH*, 13(2), 229-237. Online: https://doi.org/10.1111/j.1365-3156.2007.01994.x.

Patz J.A., Campbell-Lendrum D., Holloway T. and Foley J.A. (2005). "Impact of Regional Climate Change on Human Health", *Nature*, 438, 310-317 (17 November) (PDF Download Available). ResearchGate. Online: http://dx.doi.org/10.1038/nature04188. Accessed July 19, 2017.

Samadoulougou S., Maheu-Giroux M., Kirakoya-Samadoulougou F., De Keukeleire M., Castro M.C. and Robert A. (2014). "Multilevel and Geo-Statistical Modeling of Malaria Risk in Children of Burkina Faso", *Parasites and Vectors*, 7(1), 350. Online: https://doi.org/10.1186/1756-3305-7-350.

Teklehaimanot H.D., Schwartz J., Teklehaimanot H. and Lipsitch M. (2004). "Weather-Based Prediction of Plasmodium Falciparum Malaria in Epidemic-Prone Regions of Ethiopia II. Weather-Based Prediction Systems Perform Comparably to Early Detection Systems in Identifying Times for Interventions", *Malaria Journal*, 3(1), 44. Online: https://doi.org/10.1186/1475-2875-3-44.

Worldbank – Burkina Faso Overview (n.d.). Online: http://www.worldbank.org/en/country/burkinafaso/overview.

Yé Y., Hoshen M., Kyobutungi C., Louis V.R. and Sauerborn R. (2009). "Local Scale Prediction of Plasmodium Falciparum Malaria Transmission in an Endemic Region Using Temperature and Rainfall", *Global Health Action*, 2(November). Online: https://doi.org/10.3402/gha.v2i0.1923.

Yé Y., Louis V.R., Simboro S. and Sauerborn R. (2007). "Effect of Meteorological Factors on Clinical Malaria Risk among Children: An Assessment Using Village-Based Meteorological Stations and Community-Based Parasitological Survey", *BMC Public Health*, 7(June), 101. Online: https://doi.org/10.1186/1471-2458-7-101.

Zhou G., Minakawa N., Githeko A.K. and Yan G. (2005). "Climate Variability and Malaria Epidemics in the Highlands of East Africa", *Trends in Parasitology*, 21(2), 54-56. Online: https://doi.org/10.1016/j.pt.2004.11.002.

World Health Organization (2014). *Quantitative risk assessment of the effects of climate change on selected causes of death, 2030s and 2050s*. WHO report. Online: http://www.who.int/globalchange/publications/quantitative-risk-assessment/en.

Spreading mosquitoes: a media analysis of Italian national newspaper coverage of mosquito-borne diseases and related interventions

Paolo GIARDULLO

Post-Doctorant, University of Padova, Italia

Malaria, yellow fever, dengue, chikungunya and more recently Zika are just some health risks related to mosquitoes. Diseases such as malaria have now been eradicated in most western countries and, since eradication, most have only registered a few imported cases brought back by travellers, if any. Indeed, according to WHO statistics, malaria is mainly linked to the African and South-Eastern Asia regions (2016). Nonetheless, in recent years some outbreaks of mosquito-borne diseases have occurred in Europe and North America. In 2009-2010, the Florida Department of Health reported 93 cases of autochthonous dengue (Hayden *et al.*, 2012); patients were infected directly in the U.S. rather than importing the disease from other countries. This had not happened since 1934 (*ibid.*). In 2007, a chikungunya outbreak in Italy counted more than 250 cases (Valerio *et al.*, 2010) and tiger mosquitoes were proven to be the vector. As the scientific literature has already shown, tiger mosquitoes have adapted to new habitats. These include the general temperate climate of Italy, as well as the urban environment (*ibid.*).

These cases demonstrate that contemporary processes involving the international mobility of goods and people, arising from globalization, actively contribute to the global spread of contagion vectors. Globalization has become a main root cause for the introduction of non-endemic species of mosquitoes into temperate zones. The most celebrated example of the expansion of the distribution of mosquitoes concerns the tiger mosquito (*Aedes Albopictus*). Its eggs have been spread globally via the used tyre trade and the importation of lucky bamboo as a domestic ornament (Medlock *et al.*, 2012). The first official recorded sighting on European soil was in 1979, in Albania, and the mosquito population has boomed since the early 1990s among Mediterranean countries (*ibid.*). Since the 2007 chikungunya outbreak in Italy, mosquito management has become a public health concern (Valerio *et al.*, 2010). Mosquitoes recently gained further international attention when, on 2 February 2016, WHO Director-General Margaret Chan announced growing concern over Zika outbreaks in Brazil, confirming the migration of the virus from Polynesia and declaring the strong suspicion of a correlation between infection and microcephaly in new-born children (WHO-Media Centre, 2016). Despite denying the need for any restrictions on travel to Brazil (*ibid.*), this alarmed several western countries over the possible spread of the disease.

Managing mosquitoes lies at the intersection between the environmental and public health policy fields and is both a challenging and ambivalent issue: to disinfest an area may be an appropriate public health intervention, while it could be controversial from an environmental perspective, affecting ecosystems and modifying the urban landscape (Mieulet and Claeys, 2014). Furthermore, it brings into the debate discussion about the regularization of trans-national trade and international mobility. Similarly, the Brazilian Zika outbreak in 2016 established mosquito-borne diseases as a global public health risk. The present contribution aims to study the articulation of these issues by reconstructing the public discourse about mosquito-borne diseases in Italian daily newspapers. The narratives and metaphors used in the coverage of other health issues in the media (Ungar, 1998; Wallis and Nerlich, 2005) will be surveyed through content analysis of six leading Italian daily newspapers. In particular, we will check for the presence of metaphors already detected in other contexts (Larson *et al.*, 2005; Stuckler and McKee, 2008) and the possible link with specific events which may increase media coverage (Claeys and Mieulet, 2013; Evensen and Clarke, 2012; Roche and Muskavitch, 2003). We will compare with other infectious diseases, to examine whether it is possible to assess the relevance of mosquito-borne diseases in Italian media discourse. We will point up shifts within the public discourse and possible connections with other policy fields. The analysis will further map out the actors involved in the narration and definition of health risks connected to mosquitoes, with a particular focus on the role of experts and policy makers. In addition, we will focus on three key analytical levels:

- whether tension exists between the health and environmental fields in media discourse and if it varies across time;
- the effect of the outbreaks registered since the early 1990s by Italian health authorities, such as chikungunya (2007) and West Nile Fever (2012), on media discourse;
- if there is a significant difference between the representations of cases located in the EU and in tropical countries.

The empirical basis for our research was the TIPS project database which collects, archives and indexes Italian daily newspaper articles (Giardullo and Lorenzet, 2016). The main theoretical approach is at the crossroads of health risk communication, environmental communication and Public Communication of Science and Technology (PCST), considering the actors and the discourse about mosquito-borne diseases. This chapter will examine a period ranging from the early 1990s to the present day; the longitudinal analysis of the main Italian daily newspapers will reconstruct the entrance of mosquito-borne diseases into the public sphere. To address these issues properly, the next section will outline the key areas for a media analysis of health issues. Then in the subsequent two sections, the analysis will focus on coverage and narrative strategies. The final remarks will summarize the main findings and reflect on the presence and structure of discourse on mosquito-borne diseases in the Italian public sphere.

Environmental and health risks in the media: a state of the art of coverage, narratives and metaphors

Health issues are increasingly present in daily news. This is a tendency which has been detected in several countries since the first attempts to 'measure' science and technology coverage in the public sphere. Almost twenty years ago, Gregory and Miller (1998), in reviewing the state of the art of science coverage in British and American newspapers, stated that the general presence of science-related content could be estimated at between 2% (in the US) and 5% (in the UK) of all articles published over three decades. Within these percentages, articles about health stories represented a considerable share, almost 75% of the total (*ibid.*, 119). Such findings were later confirmed and reinforced by several other studies. Clark and Illman (2006) detected the same tendency in the main US quality newspapers within a timespan ranging from 1980 to 2000, revealing 42.6% of health-related content in their sample (*ibid.*, 505). Zinn (2010) as well as Hamilton *et al.* (2007) revealed the particular importance of health risk as contributing significantly to health stories. Similar long-term patterns, increasing across time, have been documented in other countries: Bucchi and Mazzolini (2003) revealed that, in Italy, the prevalence of health issues in the main daily newspapers has grown significantly since the late 1940s; similarly, in Brazil, studies on science news both on TV (Ramalho *et al.*, 2012) and in newspapers have revealed a marked interest by newsmakers in looking at health-related stories. Similar findings were reported in a sample of leading German newspapers (Summ and Volpers, 2016), where medicine is the second academic field represented in the national daily press. Therefore, health issues are well represented in the news: independently of the context and of the medium (press or TV), health news often represents a significant amount of the coverage of scientific issues. It is also interesting to note the strong presence of risk: indeed, health-related risk seems to be a trigger for the newsworthiness of health issues. Health risks are popular and frequently appear in newspaper sections specifically dedicated to health. Studies on cancer prevention, or on factors and lifestyles that can cause the onset of respiratory and cardiovascular diseases, significantly contribute to the total share of scientific information.

Generalizing these different findings – that health issues tend to be above the so-called "threshold of public relevance", as postulated by Lorenzet and Neresini (2014) in a competitive arena of public issues (Hilgarntner and Bosk, 1989) in which the space and coverage given to an issue is necessarily linked to the carrying capacity of the media (the pages of a newspaper for instance, or the number of reports on a TV news bulletin) –, only some issues are "strong" enough to become visible. To attain visibility means to become a story that may be covered for several days; moreover, for an issue it means to be recursive and more or less stable across time. According to Lorenzet and Neresini (2014), to overcome such a threshold an issue needs to be anchored to other elements. They built on the framing theory of the media (Scheufele, 2010) and on the social representation approach (Farr and Moscovici, 1984) to argue that an issue needs to be framed in a way to interest and be understandable for the audience, and further it must intercept the social representations that connect processes to everyday life and practices. A specific issue may enter – or even re-enter periodically – because it is connected to other issues that closely affect the interests of the public or even their practices. If an issue has an easy connection with the needs or interests of the general audience, its coverage may be more likely, stable and recursive across time. This is evident if we consider the ups and downs of climate change in the media. According to different studies, climate change entered the media public discourse quite recently even though concern over such global change began growing and the debate surrounding it started to gain momentum in the late 1980s. Ungar's analysis (1998) compared it to other complex environmental issues, such as ozone depletion in the atmosphere. During his analysis, he noted the lack of a specific link between climate change and daily life; as Ungar showed, the difficulty of connecting the effects of climate change to people's daily activities contributed to the delay in its entry into the public sphere between the 1990s and early 2000s. On the contrary, ozone depletion was quickly portrayed using the metaphor of the "hole in the atmosphere"; furthermore, it was linked to everyday life through measures such as the inclusion of a UV index in weather forecasts, making ozone depletion an "intrusive problem" (*ibid.*, 307). This lesson from environmental communication and PCST allows us to interpret in a more precise way the reasons behind the success (or the lack thereof) of a specific topic within the public sphere. It further suggests expanding the scope of the analysis well beyond the gravity of a specific threat or issue by actively looking at the broader context to understand the processes of media presence and recursion across time. Attention cycles (Downs, 1972) on specific issues may be influenced by sudden and – to the public – unexpected events such as a crisis or alarm situation (Djerf-Pierre, 2013).

Indeed, as well as being a recurrent and significant presence, the topic of health risk may become predominant in newsmakers' agendas in situations of alarm or emergencies; these are inevitably political and media events (Hodges, 2000). Exactly as with environmental disasters and accidents (Anderson, 2014), health crises linked to pandemics and epidemics receive a lot of attention from the media, even though they do not necessarily coincide with the "advent" of the emergency. Indeed, as the literature available on health crises shows, pandemics and epidemic diseases represent the classic tip of the iceberg. Partially anticipating the findings of the present contribution, this was confirmed in our research into mosquito-borne diseases in Italian daily newspapers.

Returning to the review, what is most noticeable with crises and alarm situations is the relevance for research on communication narratives in health. Real or just feared scenarios of a fatal pandemic have been extensively analysed by scholars who monitored the media or later reconstructed the entire history of a specific health risk or disease. Sheldon Ungar (1998), for instance, surveyed the top quality daily newspapers in the UK and reconstructed the media narrative surrounding the Ebola virus outbreak in Central Africa. He concentrated on the potential effects of narratives in triggering anxiety (inducing fear) or, on the contrary, in reassuring (reducing fear) about the risk of the Ebola virus reaching the UK. Similarly, Patrick Wallis and Brigitte Nerlich (2004) studied alarm over SARS by sampling British daily newspapers during the winter of 2003-2004; they concentrated on metaphors used to frame the (at that time) uncertain nature of the disease. Metaphors and the effect they may produce are extremely interesting. Indeed, they contribute to framing specific news stories by creating what Gamson and Modigliani (1989) call "interpretative packages", like for instance Ungar's processes of inducing fear / reducing fear. More recently, this concept has been translated into the term "frame" (Scheufele, 1999; Scheufele and Tewksbury, 2005). Indeed, Shih *et al.* (2008) analysed the coverage and narratives of three public health epidemics – namely BSE, West Nile Virus and Avian Flu – to show that frames can vary both across time and between issues. In this context, metaphors are among the most powerful symbolic devices which define the actual core of a narration and, moreover, they contribute to making it memorable and communicable (Ungar, 1988). So, metaphors are part of the packages used by newsmakers to frame specific events; nonetheless, as Stuckler and McKee (2005) have argued, metaphors are heavily used in public discourse directly by political actors, especially when they present opinions about or justifications for their strategic choices on health issues. Metaphors about security, even evoking militaristic comparisons ("being under siege", "the war on cancer", "to defeat the virus") have been shown to be widely adopted in public communication on health. Larson *et al.* (2005), for instance, analysed the presence of militaristic metaphors in articles talking about SARS in comparison with others about invasive species. They provided an interesting study by evidencing how the newspapers tended to personify both pathogens and invasive species. According to Larson *et al.*'s interpretation, this assigned agency to these elements and justified a militaristic response. For instance, invaders deserve to be warded off, killer viruses need to be engaged in a battle. These metaphors are often used in narratives surrounding health-related issues both by journalists and institutional actors involved in a specific health related story, be it a crisis or not. Vallis and Inayatullah (2016) reconstructed the use of metaphors for justifying health policy choices to contrast tuberculosis and obesity; in doing so they also highlighted the heuristic value of metaphors in shaping views about such issues. Vallis and Inayatullah particularly stressed the power of metaphors in promoting stigmatization processes: in the militaristic metaphor construction a relevant role is assigned to someone seen as a "collaborator" (a sufferer of a disease) who can be blamed for not actively opposing their invaders (the virus or pest). In studying SARS in the context of a feared health crisis, Eichelberger (2007) showed how the media systematically blamed Chinese migrants for propagating the contagion in New York; the blame was justified by a cultural otherness often characterized as anti-hygienic (*ibid.*, 1287).

This overview has shown how many notions and concepts have been developed through research experiences within the field of health risk communication, environmental communication and PCST. Summing up the elements that have emerged so far, it can be argued that the key principles to be kept in mind in a new health-related topic analysis, such as mosquito-borne disease, can largely draw on the literature about health crises. Therefore, the main points to deal with are the coverage of a specific issue across time and defining attention cycles, looking for possible connections with audience interests and needs; how the narratives build on the metaphors as reported in the news; and the presence of actors and how they are linked with metaphors and narratives. These elements offer enough flexibility to address the analytical points raised in the introduction to this chapter. Compared to other media studies of risks connected with mosquito-borne diseases (Evensen and Clarke, 2012; Roche and Muskavitch, 2003) focused on assessing the quality of the message from the mass media for a general audience, this contribution will offer an original overview of the public relevance assigned by the media to mosquito-borne diseases by analysing the outbreaks reported in Italy and in other parts of the world, the possible differences in describing them, and the role of specific actors such as experts and policy makers.

How to catch mosquitoes in the papers? The strategy for constructing corpora and analytical methods

To analyse mosquito-borne disease coverage in the papers this contribution will take advantage of the data provided by the TIPS monitoring project (Giardullo and Lorenzet, 2016), which is a media monitoring project established at the University of Padova that allows research to be conducted into the main Italian newspapers. The TIPS project is based on a continuous collection of online newspaper articles as a source for analysing public representations of science and technology.

Table 1. TIPS project articles available from the monitoring platform

Newspaper		Collection starting date	Articles available as of 31 December 2016
1	*Corriere della Sera*	1st January 2010	209488
2	*La Repubblica*	1st January 2010	179011
3	*Il Sole-24 Ore*	1st January 2010	213194
4	*La Stampa*	1st January 2010	187266
5	*Avvenire*	1st January 2011	51018
6	*Il Giornale*	1st January 2011	139412
7	*Il Mattino di Napoli*	1st January 2014	96988
8	*Il Messaggero*	1st January 2014	101272
Total 2010-2016		Restricted set (1-6)	788959
		All newspapers	1177649

Source: TIPS project database.

As indicated in the table above, the total number of articles available up to December 2016 is considerable; for the present analysis, the restricted set of sources (*Avvenire, Corriere della Sera, Il Giornale, Il Sole-24 Ore, La Repubblica, La Stampa*) offered the best combination of chronological and exhaustive coverage. Drawing on such a large volume of articles available for queries since 2010 (*Table 1*), it is possible to investigate topics across time. It was also possible to explore a sample ranging from 1990 to 2013 from the two most widely-read quality Italian newspapers, namely *Corriere della Sera* and *La Repubblica*; this sample was created using the constructed week methodology (Hester and Dougall, 2007). The main aim of using a sample was to expand the opportunity to investigate a longer time span, possibly exploring long-term patterns. It should also be kept in mind that the *Aedes Albopictus* was observed in Mediterranean countries at the beginning of the 1990s. The entire sample contains about 163,323 articles; the present analysis, to avoid overlaps and redundancy with the monitoring platform, considered the span from 1990 to 2009, using in total 117,707 articles.

Article selection followed a two-step path in a funnel strategy. First, two corpora were extracted using the compound query of two stemmed words "mosquito* OR invasive specie*[6]", one for the 1990-2009 sample and another for the 2010-2016 collection. Secondly, a further step in selection was applied directly to the corpora by selecting articles whose content concerned mosquito-borne diseases. The selected articles created two corpora:

- 2010-2016 consisting of 263 cases;
- 1990-2010 consisting of 34 cases.

In order to address the main points indicated in the introduction aimed at offering a depiction of mosquito-borne disease in Italian newspapers, our analysis took a two-levelled approach. First, an overview of the temporal distribution of articles and the salience of the topic, then, in the subsequent section, a mainly qualitative content analysis using the QDA-Miner CAQDAS package, in order to catch the main narratives about mosquito-borne disease and possible interventions by analysing metaphors and actors. The main categories adopted (e.g. fear reducing/inducing, militaristic metaphors, blaming) were selected according to the literature explored in the previous paragraph, keeping in mind the specific features of mosquito-borne diseases as a case study.

Mosquitoes in the papers: analysis of long and mid-term coverage

The entire collection of articles retrieved from the TIPS database was quite small. Taking into consideration the percentage of articles about the topic as a more precise measure of its salience, we can see that, on average, both corpora contained a limited number of articles about mosquito-borne diseases. In the constructed week, the percentage for mosquito-borne diseases was 0.028%; for the 2010-2016 full collection of online published articles there was a similar value (0.033%). Thus, it would be easy to conclude that the general importance assigned by Italian newspapers to this issue is extremely low. Moreover, this low importance seems to be stable across time. However, to better assess the coverage we

6 The original Italian wording is "Zanzar* OR specie invasiv*". I chose to use this compound query because of the very poor results obtained through queries with names of diseases, like for instance "chikungunya" or Zika. This more general query produced a better corpus to be explored.

need to consider the percentage attained by other issues; for instance, the general percentage of articles with techno-scientific content is much higher, reaching 7.25% in 2016. The graph below further highlights the very limited importance assigned to mosquito-borne diseases in the two macro-periods selected for the analysis.

Figure 1. *Comparison of coverage of infectious diseases in the Italian daily press for 2010-2016 (N=788959) and 1990-2009 (n=117707), % values.*

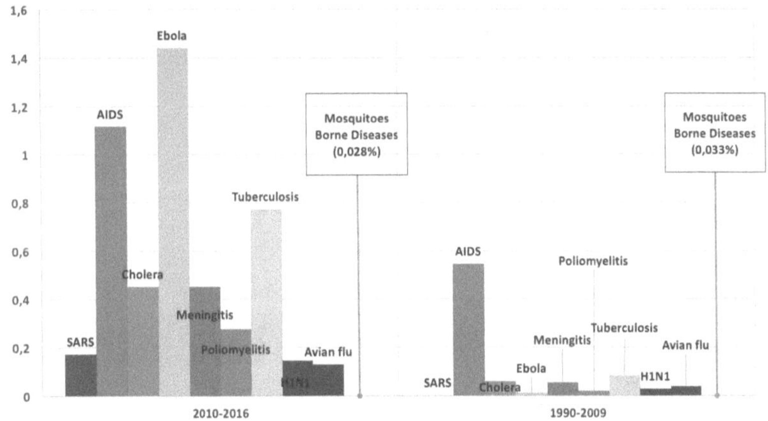

Source: Own elaboration based on the TIPS project database

Indeed, in merging articles in which there was at least one occurrence of diseases having mosquitoes as vectors,[7] their presence was definitely confirmed at the bottom of coverage of health issues compared to a selection of other infectious diseases. A first and easy interpretation of this evidence might be that mosquito-borne diseases failed across time to gain sufficient anchorage to overcome the threshold of public relevance. Even though the figures are clear in suggesting this interpretation, the two "snapshots" in *Figure 1* do not illustrate how it evolved across time. As has been mentioned before, the attention cycles theory of Downs (1972) and its further empirical applications (Djerf-Pierre, 2013) suggest that constancy across time should not be taken for granted. Ebola, for instance, seemed to be a main topic among health-related issues in the 2010-2016 period; but actually, if we split the data by yearly granularity, we can see that Ebola

7 These included Zika, Dengue, West Nile Fever, Chikungunya, Malaria, Yellow Fever, Japanese encephalitis, St. Louis encephalitis, etc.

had high coverage only in 2014, with 683 articles published, which equals about 6.1% of the yearly total coverage. In 2014 a new outbreak in the Democratic Republic of the Congo, central Africa, was registered while another one spread in different West African countries (Guinea, Liberia, Mali, Nigeria, Senegal, Sierra Leone); the death rate was not the highest ever recorded but several EU and US citizens working in the region became infected. Among them were Italian medical personnel who had volunteered in Sierra Leone who were put in quarantine once they got back to Italy,[8] bringing the threat inside national borders. In the same period a vaccine was developed and tested in an Italian lab for the emergency.[9] The Ebola example suggests the need for specific care when analysing data for the present case study. Indeed, it is worth checking the distribution of coverage of mosquito-borne diseases on the one hand; and, on the other hand, the hypothesis of crisis relevance should be considered as well.

In the sample obtained through the constructed week (1990-2009) there was no significant peak of coverage across the years; the mean of the articles was so low that it was not worthwhile to consider the seven articles published in 2005 as a media attention peak. The seven articles concentrated mainly on initiatives for raising funds for combating malaria in Africa within a specific campaign promoted in Italy.[10]

This is particularly interesting and, at the same time, striking because in the period spanning between 1990 and 2009 two key events for mosquito-borne diseases directly affected Italy. In 1990, the presence of *Aedes Albopictus* was recorded for the first time in a kindergarten in Genoa, Liguria (Sabatini *et al.*, 1990): experts signalled the possible risk by describing it as a new vector for tropical diseases in Europe,[11] although this was enough to push for coverage on this. In 2007, the aforementioned chikungunya outbreak exploded in Ravenna province, Emilia-Romagna, in north-eastern Italy, with more than two hundred infected patients (Valerio *et al.*, 2010). In 2010 two cases of dengue were also reported in Bologna (Emilia-Romagna) by a couple back from a vacation in Thailand;[12] there,

8 *La Repubblica*, Ebola: medico ricoverato a New York positivo a virus. Il primo caso nella metropoli USA, 24 October 2014.

9 *La Repubblica*, Ebola, con il vaccino sviluppato in Italia si ottiene l'immunità per 10 mesi, 10 September 2014.

10 *Corriere della Sera*, L'Africa è qui tra noi: oggi va in piazza la solidarietà, 28 May 2005.

11 *La Repubblica*, La "zanzara tigre" é sbarcata in Italia, 21 December 1990.

12 *La Repubblica*, Dopo la febbre Chikungunya altri due casi di dengue, 9 September 2010.

the reference to chikungunya was made explicit (see title in the note 7) but it did not result in a news story able to last longer.[13]

These three cases exemplify different elements which may play a significant role in pushing media coverage of a health issue – namely, the connection to other narratives (fundraising and the precarious health conditions in Africa), the sensationalism of a specific event and the possible risk for the population in Italy, so within the country's borders, something that can be called an "intra-moenia danger", like with Ebola. Nonetheless, the general trend was not influenced, as none of these elements actually led to significant coverage.

In the period spanning from 2010 to 2016 the number of articles selected through the query was seven times higher than in the 1990-2009 sample. As already pointed out, the share of articles about mosquito-borne diseases was almost the same as in the previous period, but there was a specific peak of attention, as shown in the next graph.

Figure 2. *Article coverage of mosquito-borne diseases (2010-2016) and the share of disease occurrences (n=395) within articles in 2016.*

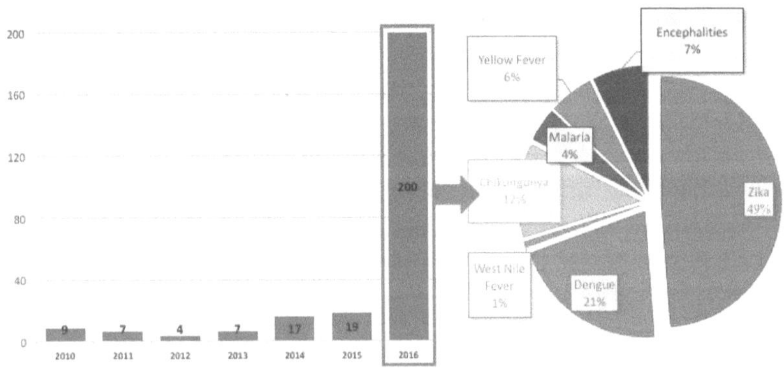

Source: Own elaboration based on the TIPS project database

[13] This lack of newsworthiness at the national level does not coincide with the local press; indeed, *Il Resto del Carlino* a newspaper mainly in circulation in Emilia-Romagna and Marche covered the chikungunya outbreak with 23 articles during late summer 2007.

The articles in the corpora were not distributed equally across time and were concentrated mainly in 2016, with an average of one article every two days published on the topic (a mean average of 16.75 articles per month). The coverage reached its apex in February 2016 with 62 articles published, then decreased until the summer when another peak became visible in August with 24 articles. Such a high concentration seems to be mainly justified by the prominence of Zika, which covered almost half of the total occurrences of mosquito-borne diseases in 2016. Going further in this quantitative exploration, the number of "Zika" occurrences registered through the corpus was 193, thus it can be argued that on average almost every article about mosquito-borne diseases in 2016 mentioned Zika.

Table 2. Distribution of mosquito-borne disease occurrences within the corpus (n=263)

	2010	2011	2012	2013	2014	2015	2016
Dengue	14	9	8	13	27	53	82
West Nile Fever	1	1	1	1	3	6	3
Chikungunya	3	1	5	--	5	2	38
Malaria	1	--	2	11	9	27	17
Yellow fever	1	--	1	5	4	1	23
Encephalitis	--	--	1	1	1	4	2
Zika	--	--	--	--	--	--	193

Source: Own elaboration based on the TIPS project database

The communicative power of Zika was further evident if we consider that it was almost absent from the Italian public discourse before 2016 (see *Table 2*), even though the virus was first isolated from humans in 1968 in Nigeria (Hayes, 2009) and several outbreaks occurred in South-East Asia, equatorial Africa and Egypt between 1947 and 2007 (*ibid.*). The risk associated with microcephaly for Zika-infected pregnant women became more evident in 2016, even though an epidemic took place in French Polynesia in 2013-2014. In early 2015 the Pan American Health Organization (PAHO) had already registered the infection spreading in Colombia (Mattar and Gonzalez, 2015). Experts made the link with previous mosquito-borne disease outbreaks, depicting an already known scenario: "Soon, [for Zika] we will have, like chikungunya, local and exponential transmission in Aedes mosquitoes, which are perfectly adapted to our environment to aerotransport arboviruses. Climate change and global warming are affecting both the natural cycle of the virus and their vectors, a situation that favors the emergence and spread of diseases" (Mattar and Gonzalez, 2015).

Indeed, as mentioned above, the alarm was launched by WHO in February, confirming concern over the infection of pregnant women in Latin America. The Italian daily press picked up on the news once again and further expanded the issue. The outbreaks registered in Colombia and Brazil since January had alerted local experts as well as international health authorities, as stated in the introduction. At this stage, it is interesting to note that in early 2016, and in January and February in particular, reference was consistently made to the 2016 Olympics that were to be held in Rio de Janeiro from 5 to 21 August. References to the imminent media and global event also continued later on, as the opening ceremony approached.

> The World Health Organization (WHO) and the Pan American Health Organization (PAHO) have appealed to pregnant women not to travel to Rio de Janeiro or in other parts of Brazil where the Zika virus is found, as part of the recommendations for the Olympic Games in Rio in 2016, to be held in August.
>
> *La Repubblica*, 13th May 2016[14]

> The Zika virus scares, and a panel of scientists have asked the World Health Organization for the postponement of the Games: 'Too many risks of contagion', they write. Experts are divided, but for now Rio 2016 is confirmed.
>
> *Corriere della Sera*, 30th May 2016

> Fear of Zika virus could seriously resize the technical level of the Olympics. In fact, the risk is that there will be a significant number of defections even in the team from the United States, the colossus that from 1992 onwards (the Soviets excelled in Seoul in 1988) has always occupied the top spot in the medals ranking [...]. But the most likely damage, for Brazil, is above all economic. The Olympics, and also the Rio Carnival, are likely to be snubbed from a tourism point of view. It should not be forgotten that the epidemic, transmitted by mosquito bites, could be the cause of rare diseases of the foetus. Meanwhile, US president Barack Obama will ask Congress to allocate $1.8 billion for treatment and prophylaxis. Too late to save the Olympics?
>
> *La Repubblica*, 8th February 2016

14 All the English translations from the original Italian were done by the author.

> A mosquito off his sleep at the Rio 2016 Games, six months after the night when, for the first time, the flame of Olympia was lit in South America. In a Brazil suffering a crisis of identity, and floundering amidst protests in the streets and labour troubles, there was no need for a further issue such as the Zika virus, caused by the bites of the *Aedes aegypti* mosquito. [...] The World Health Organization has called it a 'global health emergency', but Brazilian Minister Jaques Wagner has reassured everyone: 'No one should be afraid to come to the Games, except for pregnant women', and President Dilma Rousseff has promised maximum effort: 'The vaccine is a commitment for all. There will be a struggle house-by-house to destroy mosquito larvae', confirming the commitment of 220,000 soldiers for the deployment of repellents and the delivery of brochures.
>
> *Il Sole-24Ore*, 4th February 2016

As can be seen, the discourse about Zika epidemics and the Olympics was connected to health risk and the anxiety it brought with it not only for locals – Brazilians in this case – but also for people from other countries and for international health institutions like WHO and PAHO. Nonetheless, the general Brazilian economic situation and the political instability of the last few years was also mentioned, underlining the importance of such a global event for the host country in terms of direct outcomes and competitiveness in a globalized world. Thus, the Olympics seemed to play a double role in this narration: on the one hand, they seemed to be a dangerous crossroads for spreading the disease further, exporting it to other areas around the globe. Indeed, this is what the panel of 125 scientists feared the most – a pandemic triggered by a major global event bringing people from all around the world exactly where the hottest outbreak was active. This panel asked for a delay of the Olympics as a precaution. However, such a precaution sounded inappropriate given the consequences of such a choice for Brazil. This connects to the second role played by the Olympics in this narrative: the Games were cited as potentially being a major victim to Zika-induced panic, along with Brazil itself. According to the excerpts above, the already contested president Dilma Rousseff was facing another hard crisis during her mandate, aggravating the precarious stability of a country already affected by a relevant financial crisis. Within such a problematic context, the hypothesis of a delay or even the cancelling of the Olympics, the magnitude of economic and image-related damage were melted together. Indeed, as assumed by scholars who study sporting mega-events, the

Olympics are of crucial political, economic and cultural importance due to the size of their global audience (Hayes and Karamichas, 2011).

As a partial conclusion regarding our analysis of the coverage of mosquito-borne diseases in the Italian daily press, it can be argued that without a specific anchorage to a relevant, awaited and highly media-friendly event such as the 2016 Olympic Games in Rio de Janeiro, the issue would not have overcome the threshold of public relevance. Indeed, as evidenced in *Table 2*, talking about Zika also allowed journalists to shed some light on the general issue of mosquito-borne diseases – dengue, chikungunya, yellow fever and even malaria, which were cited only sporadically in the previous years and benefitted from a "pull effect" from Zika. However the economic and political aspects also made their contribution to the anchorage of mosquito-borne diseases as a public issue in the Italian media sphere. Zika became relevant for other topics not intuitively linkable to health concerns, such as the global economy and international relations. Indeed, it seems that this kind of 'pull effect' driven by Zika went much farther than the general discourse about health risks. These elements represent a clear peculiarity of the present case study and will be recalled in the concluding remarks. In the next section, narratives and metaphors registered in the two corpora will be analysed, highlighting the similarities as well as the differences with the general aforementioned tendencies in health risk communication in the media sphere.

Narratives and metaphors about mosquito-borne diseases

What is evident from our findings above is the role of Zika as a trigger able to foster attention on the general issue of mosquito-borne diseases in the Italian public sphere. Indeed, it can be argued that without Zika's connection to the 2016 Olympic Games in Rio, mosquito-borne diseases would have been unlikely to have overcome the public relevance threshold in the Italian media sphere. Looking more closely at the general framing of the Zika outbreak as a health communication case, most striking is that it was a completely "extra-moenia danger". Indeed, so far (early 2017), people affected by Zika, as well as new born babies with linked microcephaly, are mainly located in Latin or South America, with a few rare cases in the US, as registered by international health authorities (WHO, 2017). Compare this to other cases of alarm or health risk, like for instance the case of Ebola, where coverage was spurred by the entrance of the disease within EU countries: contagion, or suspected cases of contagion, once having entered the EU could result in a "fear inducing process". This was true in the case of Ebola, as well as for SARS (Eichelberger, 2007): in both cases, coverage was triggered by the entrance of the danger within one's own borders. Thus, the peak in coverage was tied to an "intra-moenia danger". In the case of Zika, the entire narration was about events in a geographically distant context.

Figure 3. *Some of the images provided by Italian newspapers: a) Brazilian health officers checking for possible mosquito eggs; b) Colombian health officers distributing mosquito nets to pregnant women; c) A Brazilian health worker preparing to apply pesticide in a junkyard.*

Source: images retrieved through the TIPS project database.

Even the images which accompanied the articles (*Figure 3* shows a few examples) were mostly specific to Latin-American contexts. This peculiarity, together with the pivotal role played by Zika in driving the coverage of mosquito-borne diseases, seems to shape a specific narrative of the issue or, at least, shape it differently compared to other cases. This is particularly true even compared to the framing of mosquito-borne diseases as recorded well before the beginning of 2016. In the past, narratives were more like the general framing described in the literature on health risk crises and on invasive species. In particular, within the 1990-2009 corpus, it was easier to find metaphors referring to an invasion, assault, or siege by mosquitoes.

> The long war on the tiger-mosquito. The Monte Spaccato (City of Rome – Ed.) tiger-mosquito emergency continues. Many citizens phoned the office yesterday complaining that they needed to ask for medical care because of the countless insect bites on legs and arms.
>
> *La Repubblica*, 19th August 1998

Brochures and ovitraps. Mosquito hunt begins. [...] The 'tiger hunt' is open but the fight against mosquitoes is for all species [...] and since April already technicians coordinated by the Environment Protection Sector of the town [Turin – Ed.] have set up a network control consisting of one hundred and fifty scattered ovitraps on the territory, to monitor, control and restrict the larval outbreaks, which are then treated regularly [...].

La Repubblica, 22nd June 2005

Mosquitoes, the struggle is already learning. Quadrifoglio (Florence environmental utility – Ed.) declares war on mosquitoes. As of tomorrow and until 31 October, the disinfestation campaign promoted by the Florentine environmental agency will affect the public areas of Florence, Calenzano, Campio, Bisenzio, Scandicci, Sesto Fiorentino, and Signa. In cities, the fight plan will cover 60,000 wells, which will be treated with 360,000 tablets of active larvicidal principle.

La Repubblica, 4th May 2008

Looking at the extracts above, it is clear how common militaristic jargon was employed: the framing of clashing, of the need to fight back against an invasion, persisted across time and even about different episodes of disinfestation across Italy (in the Turin, Florence and Rome areas). This framing was even shared across time, during which interventions were described as manoeuvres against invaders or as a hunt to kill wild animals. Such metaphors were not only linked to local interventions by municipalities to remove the opportunity for mosquito proliferation, however; indeed, they also pertained to other domains. The following extract for instance comes from an article dedicated to a nature-conservation threat brought by mosquitoes.

Mosquito killer along with tourists. Darwin's paradise at risk. Published in an authoritative British scientific journal, rather than a zoology study it looks like an investigation by the border police. For one year, from September 2006 to October 2007, researchers from the University of Leeds and the Zoological Society of London have regularly raided the aircraft landing in Baltra and San Cristobal, the only two airports in the Galapagos Islands. [...] To undermine the priceless treasure of endemic plants and animals used by Charles Darwin as an open-air laboratory for developing the evolution theory, the mosquito that transmits malaria, avian smallpox and West Nile Fever would be especially able. [...] And while the stakes in the beautiful Pacific islands are really extraordinary, we should not underestimate the threat posed by alien species for a more 'mundane habitat' like Europe.

La Repubblica, 12th September 2009

The risk for endemic flora and fauna in the Galapagos Islands, eastern Pacific Ocean, was taken as a paradigmatic example of the threat that mosquito-borne diseases pose to environmental protection. The threat for human health was almost absent and the menace concerned the great historical and cultural value of a far-off and exotic land. Therefore, the mosquito menace was represented as a threat to the environment, for the priceless biosphere of the Galapagos Islands, an area to be preserved for its unique biodiversity. Nonetheless, in the conclusion of the article the narrative turns reflexively towards a closer setting like Europe. A less exotic context – at least from the point of view of the journalist – could be endangered by an "invasion". This specific use of the narrative, showing concern for an endangered environment because of an invasive species, was also present in the 2010-2016 corpus. In this case, it was not an autonomous narrative; rather, it was linked to the more general – and much more covered – topic of climate change and global warming. Articles about *Aedes albopictus* spreading in the Balkans and in other temperate areas of Europe consider mosquitoes as the tangible signal of a worrying temperature increase. This may suggest a similarity between the health and environmental domains in the narrating of mosquitoes as an invasive species. The presence of shared metaphors in both cases, as exemplified above, may indicate a shared semantic of the risk brought by mosquitoes through the possible solution (reclaim) and the monitoring of their presence (checking points of access and controlling places for proliferation). However, given the very limited number of articles before 2016, it is hard to explore whether or not tension exists between the health and environmental fields in media discourse. The articles were quite limited and generally linked to some local, episodic news event which was more likely to be framed as an urban governance issue for the management of annoying pests rather than as a more consistent narration on mosquito management.

It is even hard to find specific reference to blaming or stigmatizing processes in relation to specific social groups. Indeed, as Vallis and Inayatullah have pointed out (2016), stigmatization is more likely to happen when a clear emergency or a specific risk is recognized; similarly, Eichelberger (2007) detected blame for provoking the spread of disease in the case of a – feared – pandemic. This was not the case before 2016, given that mosquito-borne diseases were not a proper issue in Italian newspapers and, surprisingly, were not afterwards either. The "pull effect" of Zika did not stimulate any of the typical narratives about blaming or stigmatization. Perhaps the absence of an "intra-moenia danger" or threat did not allow the development of such a narrative in Italian daily newspapers, which chose to privilege the image of a plague hitting an already suffering country like Brazil.

Despite the emergence of Zika as an emergency, which helped lift mosquito-borne disease above the threshold of public relevance, it did not produce a typical framing of the emergency. During our research, this was further encountered when looking at the dynamics of "fear inducing/ fear reducing" (Ungar 1998): after a fear-inducing event, that is to say the boom of the Zika crisis, the media tended to tune down and offer more reassuring scenarios, but it was not a properly autonomous narrative.

> Once upon a time evolution. Today there is CRISPR, the system that allows you to change the DNA of microbes, plants, animals and humans in a few days and with a hundred euros. […] Then using a technique called 'drive gene technology' (which allows you to extend the mutations of CRISPR to both pairs of chromosomes of an individual), the altered gene extends to the entire species in a few generations. The mind – especially with the outbreak of Zika expanding – goes immediately to mosquitoes. Erasing them from the face of the planet has now become technically possible. But with what consequences for the ecosystem? For the moment, scientists have decided to stop, in front of an irrevocable decision that would have exposed them to accusations of 'playing God'.
>
> *La Repubblica*, 22nd September 2016

> The killer of malaria has the face of Valentino Gantz. 32 years old, he is the biologist who had the intuition that could eradicate forever the devastating disease. How? With a 'lab trick': a modification of the mosquito carrying the virus gene so as to create a transgenic species harmless to humans. […] Along with Ethan Bier, Gantz has devised a method that allows the introduction in a group of mosquitoes of a 'genetic kit' that makes them resistant to the parasite Plasmodium, responsible for malaria. The process renders harmless mosquitoes and refractory disease. It is a decisive step forward to stop the spread of the virus and soon they would like to test in the laboratory the scope of the discovery for other diseases such as dengue or Zika.
>
> *Il Giornale*, 25th July 2016

The fear inducing spiral started by depicting the threat to the forthcoming 2016 Olympics as a hot spot for contagion; then, the fear reducing process about Zika converged on the wider narration of molecular engineering and genetic editing. Indeed, almost analogously with the beginning of the alarm, the fear-reducing process largely built on a wider and more stable narration. Synthetic biology and gene editing represent the new frontier of biotechnologies, further expanding the scope and application of techniques for genetically modified organisms. That is to say, it is an already publicly relevant issue, which entered the public sphere through a public controversy almost two decades ago (Bauer *et al.*, 1998); moreover, as has been shown, "red" biotechnologies (for health) tend to be favoured and considered less ethically controversial than agro-biotech or so called "green" ones (Bauer, 2005). The quotations above testify to the configuration of fear-reducing patterns through scientific research outcomes. This is quite familiar in relation to other effects of media reassurance for alarms: experts as well as health authorities become dominant actors in the media narration by showing strategies for combating the emergency. Promising scenarios of a future without disease or without tedious pests cheer up the audience. In terms of metaphors, the use of violent metaphors (to eradicate, to erase, to kill), seems to reassure the audience that the advancement of biotechnology is the way to defeat the threat.

Conclusion

The exploration of mosquito-borne diseases in the Italian public sphere has shown that the issue has only recently overcome the public relevance threshold at the national scale. Using categories coming from health communication, environmental communication and PCST debates, this chapter has contributed to this volume by exploring the coverage and the narratives about mosquito-borne diseases which have recently affected the Italian public sphere. Nonetheless, the issues failed to become proper news until 2016. Through the analysis of media trends and the content of the articles, proof has been found of the close connection between the media coverage of mosquito-borne diseases and one of the most media represented events, that is to say the Olympic Games.

Zika's "pull effect" for the general issue is particularly relevant because it was definitely evident, especially compared to the chikungunya outbreak – which was a concrete "intra-moenia" outbreak of a mosquito-borne disease – but which, perhaps because of the absence of a specific or stronger anchorage, did not contribute to fostering coverage of the general issue. This offers some interesting elements for future comparison: just to summarize the most relevant of them, more attention was given to a danger still outside the Italian borders than to other past ones within Italian borders (the chikungunya and West-Nile Fever outbreaks respectively in 2007 and 2012); and the use of militaristic metaphors was more linked to fear-reducing processes, in a converging trajectory of framing with scientific research.

The main finding of the Zika pull effect offers a two-fold conclusion. On the one hand, this case confirmed the need to anchor mosquito-borne diseases to more complex issues which lie at the crossroads between health and the environment. As our research showed, without a specific anchorage able to provide a sharable frame, issues are unlikely to overcome the threshold of public relevance. This is particularly evident here: all the elements of the narrative able to depict the risky implications both for human health and the environment were always there but they did not turn the issue into a news story until the Zika virus was unintentionally and unexpectedly linked to the Olympics. On the other hand, this finding indirectly echoes the conclusion of Claeys and Mieulet (2013) about the level of concern over diseases such as dengue among inhabitants on the French Riviera: anxiety about outbreaks caused by a disease vector in their area is completely overwhelmed by the discomfort of a tedious pest. Similarly, in Italian newspapers the topic is not a proper issue and even when it is covered the diseases are rarely mentioned; this has been further confirmed by articles which consider *Aedes albopictus* and other mosquito species as a pest rather than a disease vector.

Such a connection, even though weak, seems to reaffirm once again the validity of the framing theory for the media as developed by Scheufele (2010) and the idea of media carrying capacity (Hilgartner and Bosk, 1988): newspapers are interdependent with the audience to which they refer and within which they are based. Needing to select events to put in the news, newspapers take into consideration issues which can be more easily anchored to the audience's interests. This means that coverage and the way in which topics are framed cannot stray that far from the general feelings of the whole society to which the news is referring. In other words, a context (like in the Mediterranean area of Europe) in which there is no formal national concern over mosquito-borne disease, can only produce an "unconcerned media narration". Interventions, strategies and solutions, even when specifically targeted to such issues, are not able to foster specific attention by the media like other topics do.

One critical interpretation of the major findings from this chapter is the almost total absence of public debate in Italy on this issue. It sounds rather paradoxical that so common and daily a problem (a pest) has not generated any discussion at the national scale despite the two major outbreaks of tropical diseases that have occurred in Italy. In the meantime, chemical treatments for mosquito control are still more common than other solutions for controlling vectors for such diseases, requiring public economic efforts to disseminate tablets of larvicidal substances mainly from municipalities. Moreover, it is not uncommon to find specific recommendations for limiting mosquito proliferation (e.g. do not leave water stagnating). The implications for limiting possible contagion are as such definitely linked to the control and sanitization of the territory both in urban and rural areas. Given the complex features of the mosquitoes, this cannot be demanded only of public bodies but requires citizens to be directly engaged with their own everyday environment, whether it is a densely inhabited neighbourhood in Rome or the flatlands of the Po Valley. In this sense, turning to a so-called biological citizenship (Pellizzoni, 2015), citizens should not only be committed to engaging responsibly with their own behaviour but should be able and interested in assessing that of others (*ibid.*, 172). This is becoming urgent because of the rate at which such threats for health are occurring and because of the limits of the chemical treatments mentioned above. Indeed, there is some evidence in the scientific literature that mosquitoes are developing resistance to them and thus that research is required to overcome such resistance (Samuel *et al.*, 2016). Therefore, even the present efficacy of the combination of public and citizens' efforts cannot be assured. Nonetheless, these as well as other problems linked to mosquito management do not seem to stimulate any kind of public debate in Italian newspapers at the national level. Or, more blatantly, it is possible that the issue is rarely considered and sporadically problematized in terms of its consequences for health and the environment because it has become normalized. To corroborate this argument on the topic of mosquito-borne disease, further research is needed. Comparative research at European scale could be one leg; a second leg might be about long-term influence of Zika pull effect taking into consideration recent events occurred in Italy. Indeed, mosquito-borne diseases raised up in the headlines again in September 2017: a four years child died in Brescia for malaria[15] and a new outbreak of chikungunya was registered in Lazio counting almost fifty cases[16]. Both comparative research as well as the update of the present research are

15 *Corriere della Sera*, Morire di malaria in Trentino, 7 September 2017.

16 *Il Sole-24 Ore*. Ad Anzio casi di Chikungunya: stop alle donazioni di sangue, 5 September 2017.

useful perspectives go beyond the main focus of our contribution here. Nonetheless it would be an appropriate follow-up for further research into this intriguing topic.

References

ANDERSON A. (2014). *Media, environment and the network society*. Basingstoke, Palgrave Macmillan.

BAUER M.W., DURANT J. and GASKELL G. (eds.) (1998). *Biotechnology in the public sphere: a European sourcebook*. NMSI Trading Ltd.

BAUER M.W. (2005). "Distinguishing red and green biotechnology: cultivation effects of the elite press", *International Journal of Public Opinion Research*, 17(1), 63-89.

BUCCHI M. and MAZZOLINI R.G. (2003). "Big science, little news: science coverage in the Italian daily press, 1946-1997", *Public Understanding of Science*, 12(1), 7-24.

CLAEYS C. and MIEULET E. (2013). "The spread of Asian tiger mosquitoes and related health risks along the French Riviera: An analysis of reactions and concerns amongst the local population", *International Review of Social Research*, 3(2), 151-173.

CLARK F. and ILLMAN D. (2006), "A Longitudinal Study of the New York Times Science Times Section", *Science Communication*, 27(4), 496-513. doi:10.1177/1075547006288010

DJERF-PIERRE M. (2013). "Green metacycles of attention: Reassessing the attention cycles of environmental news reporting 1961–2010", *Public Understanding of Science*, 22(4), 495-512.

DOWNS A. (1972), "Up and down with the ecology: the 'Issue Attention Cycle'", *Public Interest*, 28, 38-50.

HESTER J.B. and DOUGALL E. (2007). "The efficiency of constructed week sampling for content analysis of online news", *Journalism and Mass Communication Quarterly*, 84(4), 811-824.

ELMER C., BADENSCHIER F. and WORMER H. (2008) "Science for everybody? How the coverage of research issues in German newspapers has increased dramatically", *Journalism and Mass Communication Quarterly* 85(4), 878-893.

Eichelberger L. (2007). "SARS and New York's Chinatown: the politics of risk and blame during an epidemic of fear", *Social Science and Medicine*, 65(6), 1284-1295.

Evensen D.T. and Clarke C.E. (2012). "Efficacy information in media coverage of infectious disease risks: an ill predicament?", *Science Communication*, 34(3), 392-418.

Farr R.M. and Moscovici S. (eds.) (1984). *Social Representations*, Cambridge, Cambridge University Press.

Gamson W.A. and Modigliani A. (1989). "Media discourse and public opinion on nuclear power: A constructionist approach", *American Journal of Sociology*, 95(1), 1-37.

Giardullo P. and Lorenzet A. (2016), "Techno-scientific Issues in the Public Sphere (TIPS)", *EASST Review*, 35(4 December), 14-17.

Hamilton C., Adolphs S. and Nerlich B. (2007). "The meanings of 'risk': A view from corpus linguistics", *Discourse and Society*, 18 (2), 163-181.

Hayden M.H., Cavanaugh J.L., Tittel C., Butterworth M., Haenchen S., Dickinson K., … and Ernst K.C. (2015). "Post outbreak review: Dengue preparedness and response in Key West, Florida". *The American Journal of Tropical Mdicine and Hygiene*, 93(2), 397-400.

Hayes E.B. (2009). "Zika Virus Outside Africa", *Emerging Infectious Disease Journal-CDC*, 15(9), 1347–1350. doi:10.3201/eid1509.090442

Hayes G. and Karamichas J. (eds.) (2011). *Olympic Games, mega-events and civil societies: Globalization, environment, resistance*. Basingstoke, Palgrave-Macmillan.

Hodges A. (2000). "Emergency risk management", *Risk Management*, 2(4), 7-18.

Larson B.M., Nerlich B. and Wallis P. (2005). "Metaphors and biorisks the war on infectious diseases and invasive species", *Science Communication*, 26(3), 243-268.

Mattar V. and González T. (2015). "Now is the time for the Zika virus", *Revista MVZ Córdoba*, 20(2), 4511-4512.

Massarani L., Buys B., Amorim L.H. and Veneu F. (2005). "Science Journalism in Latin America: A case study of seven newspapers in the region", *Journal of Science Communication*, 4, 3.

Medlock J.M., Hansford K.M., Schaffner F., Versteirt V., Hendrickx G., Zeller H. and Van Bortel W. (2012). "A review of the invasive mosquitoes in Europe: Ecology, public health risks, and control options", *Vector-Borne and Zoonotic Diseases*, 12(6), 435-447. doi:10.1089/vbz.2011.0814

Mieulet E. and Claeys C. (2014), "The implementation and reception of policies for preventing dengue fever epidemics: a comparative study of Martinique and French Guyana", *Health, Risk and Society*, 16(7-8), 581-599. doi:10.1080/13698575.2014.949224

Pellizzoni L. (2015). *Ontological Politics in a Disposable World: The New Mastery of Nature*. Farnham, Ashgate Publishing Ltd.

Ramalho M., Polino C. and Massarani L. (2012). "From the laboratory to prime time: science coverage in the main Brazilian TV newscast", *Journal of Science Communication*, 11(2), A02.

Roche J.P. and Muskavitch M.A. (2003). "Limited precision in print media communication of West Nile virus risks", *Science Communication*, 24(3), 353-365.

Romi R., Severini F. and Toma L. (2006). "Cold acclimation and overwintering of female Aedes albopictus in Roma", *Journal of the American Mosquito Control Association*, 22(1), 149-151.

Sabatini A., Raineri V., Trovato G. and Coluzzi M. (1990). "Aedes albopictus in Italy and possible diffusion of the species into the Mediterranean area", *Parassitologia*, 32(3), 301-304.

Samuel M., Oliver S.V., Coetzee M. and Brooke B.D. (2016). "The larvicidal effects of black pepper (*Piper nigrum L.*) and piperine against insecticide resistant and susceptible strains of Anopheles malaria vector mosquitoes", *Parasites and Vectors*, 9(1), 238.

Scheufele D.A. and Tewksbury D. (2007). "Framing, agenda setting, and priming: The evolution of three media effects models", *Journal of Communication*, 57(1), 9-20.

Scheufele D.A. (2010). "Framing as a Theory of Media Effects", *Journal of Communication*, 49(1), 103-122.

Summ A. and Volpers A.M. (2016). "What's science? Where's science? Science journalism in German print media", *Public Understanding of Science*, 25(7), 775-790.

Stuckler D. and McKee M. (2008). "Five metaphors about global-health policy", *The Lancet*, 372(9633), 95-97.

UNGAR S. (1992). "The rise and the (relative) decline of gloabl warming as a social problem", *Sociological Quarterly*, 33, 483-301.

UNGAR S. (1998). "Hot crises and media reassurance: acomparison of emerging diseases and Ebola Zaire", *The British Journal of Sociology*, 49(1), 36-56.

UNGAR S. (2000). "Knowledge, ignorance and popular culture: climate change versus the ozone hole", *Public Understanding of Science*, 9, 297-312.

VALERIO L., MARINI F., BONGIORNO G., FACCHINELLI L., POMBI M., CAPUTO B., … and DELLA TORRE A. (2010). "Host-feeding patterns of Aedes albopictus (Diptera: Culicidae) in urban and rural contexts within Rome province, Italy", *Vector-Borne and Zoonotic Diseases*, 10(3), 291-294.

VALLIS R. and INAYATULLAH S. (2016). "Policy metaphors: From the tuberculosis crusade to the obesity apocalypse", *Futures*, 84, 133-144.

WALLIS P. and NERLICH B. (2005). "Disease metaphors in new epidemics: the UK media framing of the 2003 SARS epidemic", *Social Science and Medicine*, 60(11), 2629-2639.

World Health Organization (2016). *World health statistics 2016: monitoring health for the SDGs, sustainable development goals.*

World Health Organization (2017). *Zika situation report, 2nd February 2017.* Online: http://www.who.int/emergencies/zika-virus/situation-report/2-february-2017/en/. Accessed 27th February 2017.

ZINN J.O. (2010a). "Risk as discourse: Interdisciplinary perspectives", *Critical Approaches to Discourse Analysis across Disciplines Journal*, 4 (2), 106-124.

Comfort-based mosquito control and vector control in the context of socio-environmental change: French experience on both sides of the Atlantic

Cécilia CLAEYS

Associate Professor, Aix-Marseille University, LPED

Mosquitoes have been the target of management initiatives in both mainland France and the French Antilles for decades, but mosquito-control policies in the two territories do not share the same origins. Mosquito control policies in the French Antilles were initially health-related and remain so today. Vector control (VC) strategies are employed to eliminate the mosquitoes that vector diseases. In mainland France, on the other hand, large-scale public mosquito-control was initiated to reduce a nuisance.

The neat separation between vector control in the French Antilles and comfort-based mosquito-control in mainland France has nevertheless been challenged in recent years by several socio-environmental changes. In the French Antilles, classic insecticide treatments are no longer effective for curbing the proliferation of *Aedes aegypti*, thus exposing the population to recurring epidemics of dengue fever and, more recently, chikungunya and Zika virus. In mainland France, the classic comfort-based mosquito control methods are powerless in the face of the recent introduction of *Aedes albopictus*, which is also a vector of dengue fever, chikungunya and Zika virus. Originally from South-East Asia, *Ae. albopictus* actually resembles *Ae. aegypti*, originally from Africa, in many respects. In addition to their similar black and white stripes, both species are classified as "domestic" by entomologists (Chan *et al.*, 1971; Salvan and Mouchet, 1994). They also share the same attraction to anthropic larval habitats and both proliferate in urban and peri-urban zones. Both species also vector several arboviruses. There is an approved and efficient vaccine for yellow fever, but not for dengue, chikungunya or Zika virus.[17] While *Ae. aegypti* tends to remain in tropical zones, *Ae. albopictus* can tolerate temperate climates and even continental climates with colder winters (Hanson and Graing, 1994). The two species are both vectors and a nuisance. Their fondness for urban and peri-urban spaces makes humans privileged prey for females on the lookout for the blood meal necessary for egg maturation.

These two mosquitoes and the arboviruses they vector have evolved across time and space. In doing so, they have overcome the physical borders and immaterial boundaries of our societies, themselves under transformation: the boundaries between wild and domestic, between mainland and overseas France, between VC and comfort-based mosquito control, between health and the environment. Given this, this chapter will analyse the evolution of public mosquito-control policies in the context of the socio-environmental changes of which they are both a cause and consequence.

We will begin by presenting our materials and methods, then we will recall the origins of mosquito-control policies and their contemporary evolution in mainland France and the French Antilles. Lastly, we will examine the processes of socio-political and socio-technical arbitration between environmental and health issues conducted on both sides of the Atlantic and ultimately examine the resistance of post-colonial taxonomies.

Fieldwork and methodology: long-term sociological monitoring

Most maps of France show what is generally still referred to as mainland France, the hexagonal form located on the eastern side of the Atlantic Ocean. Some maps also include a few pieces of confetti grouped together in a sidebar. Two of these are the French Antilles: Guadeloupe, with its butterfly shape, and Martinique, the smaller of the two, with the knobby protuberance on its eastern flank formed by the Caravelle peninsula. In this sidebar, they are next to French Guiana, Reunion and Mayotte. These far-off islands placed side-by-side in the same box to

---

17 See Fouque (*infra*).

optimize map space are legacies of colonial France. Spread throughout the Caribbean and Indian Ocean, these islands are now called DROMs: Overseas Departments and Regions. Administratively, they are French – and therefore European. And yet, DROMs have historical, cultural, socio-economic, climatic and geographical specificities that set them apart from mainland France more than they unify them. Moreover, the term DROM tends to mask the internal diversity of these territories, further fallout from an ethnocentrism that collates overseas territories that actually have diverse historical legacies and contemporary realities (Vergès, 2006). Given this, our comparison between mainland France and the French Antilles is by no means an overview of all DROMs.

The analyses presented here are based on sociological monitoring of a diverse corpus (interviews, questionnaires, direct observation, press clippings) over a twenty-two-year period. The corpus of data was compiled in the context of two sociology theses (Claeys-Mekdade, 2000; Mieulet, 2015), two European research programmes – life (Claeys-Mekdade and Nicolas, 2002) and life+ (Claeys and Mieulet, 2013), a national research programme (Claeys *et al.*, 2015) and local research contracts (Claeys and Morales, 2002; Claeys *et al.*, 2009; Claeys and Mieulet, 2012). The geographical and thematic scope of this research has broadened over time based on the evolution of the topics and the controversies that have surrounded them. The surveys in mainland France were initially focused on the Rhône Delta, but were later extended to the entire Mediterranean coast and South Corsica. In overseas France, the most in-depth surveys have focused on Martinique and Guadeloupe.

The analyses presented here are based on a selection of semi-structured interviews and questionnaires from different research initiatives. A first set of surveys was conducted between 1995 and 2001 and focused on the Camargue region. This included fifty-seven semi-structured interviews, monitoring through direct observation of public consultation meetings and expert debates, as well as a questionnaire-based survey with 260 inhabitants in the municipalities of Arles, Saintes-Maries-de-la-Mer and Aigues-Mortes. The survey campaign conducted between 2009 and 2012 covered the entire French Mediterranean coast and South Corsica. It included fifteen formal and informal interviews with institutional actors and field operators involved in the implementation of mosquito-control policies and health prevention measures, healthcare professionals and 149 semi-structured interviews with local inhabitants. In 2010, a questionnaire-based survey was conducted with 281 inhabitants in the

Alpes-Maritimes. In 2012, a new campaign involving questionnaire-based surveys focused on 932 inhabitants along the entire French Mediterranean coast, including 140 inhabitants in the Alpes-Maritimes who had previously been interviewed in 2010. Direct observation was conducted when sociologists joined mosquito-control teams as they intervened in people's homes in the Alpes Maritimes in 2009 and in Martinique and Guadeloupe in 2012. Direct observation techniques were also used to monitor annual reporting meetings at the Direction Générale de la Santé (Health Department) and the Agences Régionales de Santé (Regional Health Agencies), as well as during conferences that brought together institutional actors, mosquito-control agents and scientists in mainland France and the French Antilles. In 2014, 160 semi-structured interviews were conducted with inhabitants in neighbourhoods comprised of single-family homes in mainland France (Nice and Marseille) and in the French Antilles (Petit-Bourg in Guadeloupe and Le Vauclin in Martinique).

All the semi-structured interviews used here were entirely transcribed and underwent standard manual thematic analysis. The data from the questionnaires underwent descriptive statistical processing and the correlations mentioned are significant based on Pearson's chi-squared test.

Prevention in the South, Comfort in the North...

The colonial origins of overseas vector control

History books on the French Antilles underscore the extent to which fevers decimated the first European colonists and soldiers to settle in these tropical regions (Abenon, 1992; Buttel, 2007; Sainton *et al.*, 2015). Such fevers were well known to the Carib population who avoided inhabiting areas near marshlands.[18] Conversely, the first colonists built their forts and cities specifically in palustrine zones (Fort-de-France in Martinique and Point-à-Pitre in Guadeloupe), a choice driven by military and commercial strategies. The best shelter from the sea for large European caravels was tucked well inside the bays most protected from bad weather and military attacks, but which were also extremely wet and infested with vector mosquitoes. As coastal navigators, the Carib peoples did not face

[18] During the Spanish colonization of South America, it was based on the native pharmacopeia that the Jesuits "discovered" the medicinal properties of the bark of *Cinchona officinalis*, from which quinine was later extracted.

the same constraints, since they could hoist their light vessels directly onto shore.

Fewer authors mention the role of European colonization in the introduction to the Caribbean of two additional vector mosquito species. Frédéric Darriet (2014) has shown how triangular trade allowed *Ae. aegypti* (a vector of yellow fever, dengue, chikungunya and Zika) and *A. Gambiae* (a malaria vector) to cross the Atlantic. Eggs and larvae made the crossing in the water reserve on ships. The insects continued their life cycle on board, using the crew and slaves for the blood meals necessary for reproduction. In addition to the dire human and health-related consequences, the slave trade also appears to have increased and diversified the risk of vector-borne disease in the Caribbean and Americas. Given the point to which it has adapted and ultimately favoured anthropic areas, *Ae. aegypti* is classified by entomologists as a domestic mosquito. It lays its eggs, finds food and reproduces in such areas. *A. Gambiae* prefers wetlands and settled along the coastal marshes in the French Antilles.

Unlike mainland France,[19] the French Antilles were never the focus of largescale public policies aimed at draining and "cleaning-up" the marshlands. The filling of the marshes in the bays of Point-à-Pitre and Fort-de-France began with the disparate piling up of urban materials and waste, which actually encouraged the formation of larval habitats. Due to a lack of means, the Public Health and Hygiene Councils established on 18 December 1848 were inefficient against the malaria and yellow fever epidemics that regularly ravaged the French Antilles. It was only after the prestigious "France" passenger ship returned to mainland France with its crew afflicted by a yellow fever epidemic contracted in the Antilles that the centralized French government truly took the issue seriously. Following this incident, a Health and Prevention Department was created in December 1908 and a Hygiene and Microbiology Institute was founded in 1910. The report delivered in 1908 by a scientific committee was unambiguous: the main cause of yellow fever epidemics was the proliferation of mosquitoes encouraged by urban insalubrity. The authors insisted on the role played by the "poor state of watermains, the poor distribution of drinking water in the city, the existence of drainage wells in the streets, the permanent opening of most of these wells, the stocking

19 Particularly in the context of large planning projects undertaken under Napoleon III, such as coastal dykes, the planting of the Landes forest and the embankment of the Rhône River.

of rainwater in tubs and barrels in yards, gardens and outbuildings around homes" (quoted by Icheck 2004, translated here).

In the following decades, the activities of the hygiene department focused mainly on larval habitats around people's homes and very little on public equipment. Analysis of the local press at the time conducted by Vanessa Icheck (2004) shows the great unpopularity of the "disinfection brigades" from the Prevention Department. They were criticized for the intrusive and impolite nature of their interventions and for their unsuccessful results in curbing the proliferation of mosquitoes. This discontent appears to have been exacerbated by a sense of unequal treatment. Among the measures taken to protect against epidemics, European civil servants received a bonus to enable them to live in districts that were higher up and farther from the insalubrious lowlands. The lowlands as such remained primarily inhabited by poor populations of colour, thus reinforcing a historical process whose origins were in part rooted in the (non-)management of the consequences of abolishing slavery (Schoelcher decree of 27 April 1848). Indeed, while a non-negligible share of newly freed slaves agreed to remain on the plantations in exchange for a relatively unattractive work contract, others opted to settle in towns and villages. Some were able to purchase a plot of land to farm (Buttel, 2007), while others moved onto vacant land, ravines and coastal areas without right or entitlement (Brissac, 2011). The protection of these Creole populations stigmatized for their poverty was not a budgetary priority for health authorities directed by and for white Europeans (wrongly) convinced that black and mixed-race inhabitants were immune and/or resistant to the fevers (Icheck, 2004).

Veritable urban sanitation projects were only implemented much later. In Fort-de-France, for example, renovation work on the waterworks and wastewater networks, whose improvement was assigned to engineer Gandillon, only started in 1933, and the efficiency of this work was limited (Icheck, 2004). In 1943, a Pasteur Institute report once again criticized the "poor state of the distribution network and its deplorable working conditions [...] a legacy of many long years of negligence in its maintenance and working" (quoted by Icheck 2004, translated here). Alongside this, health authorities initiated chemoprophylaxis campaigns. From 1934 to 1949, schoolchildren were treated in the municipalities most affected by malaria. Initially, quinine doses were distributed, followed by synthetic drug formulations (quinacrine, premaline and plasmochine, notably).

The 1950s marked a new turn with the development of insecticide treatments. DDT (Dichlorodiphenyltrichloroethane), followed by HCH (Hexachlorocyclohexane) were used in spray campaigns, but also in indoor residual spraying based on recommendations from the World Health Organization (WHO). Then, these organochlorines were replaced by organophosphorus insecticides. They were effective to a certain extent against epidemic risks. But, the abundant and regular use of organochlorines, followed by organophosphorus treatments resulted in the genetic mutation of mosquitoes in these insular environments. The insects developed strong resistance which resulted in their renewed proliferation (Darriet, 2014; Marcombe *et al.*, 2009).

And just as the mosquito vectors were developing strategies to resist insecticides, the human population saw its vulnerability increase too. Indeed, the successive agricultural crises and resulting rural exoduses, which culminated with the decline of the sugar industry in the late 20th century led new and poor populations to occupy coastal areas, concentrating into towns or spreading along the coast (Burner, 2015). This demographic history is now visible in the general pattern of land use in the Antilles. Regarding Guadeloupe, Brissac (2011) has highlighted the following tendencies: rich neighbourhoods are mainly located on hills overlooking the sea; middle class districts are mainly in the intermediary zone; and poor districts are concentrated along the coast. Such anarchic urbanization exposed an even larger number of inhabitants to the proliferation of *Aedes aegypti*, which thrives in anthropized coastal areas.

The last indigenous cases of yellow fever in the French Antilles occurred in the 1950s and the last cases of malaria in the 1960s (Icheck, 2004). There is now a vaccine against yellow fever.[20] Largescale modernization initiatives in the capital cities (the Fort-de-France airport in the Abysses district opened in 1950, construction of the Pointe-à-Pitre port and commercial zone in the Jarry district undertaken in 1965) resulted in the filling of the large marshes so conducive to the mosquitoes that vector malaria. Dengue fever transmitted by *Aedes aegypti* did not disappear and even increased significantly starting in the 1980s. The introduction of the chikungunya virus in 2014 and of Zika virus in 2015 for which *Aedes aegypti* has vector competence rapidly resulted in new epidemics.

---

20 The yellow fever vaccine was finalized in 1932, approved by WHO in 1948 and has been gradually disseminated ever since.

The invention of comfort-based mosquito control

As Frédéric Darriet (2014) has explained, there is a long list of regions in mainland France historically affected by malaria: Camargue and the Mediterranean coastal lagoons, Corsica, Morbihan, the Marais Poitevin, Sologne, Strasbourg and its surroundings, the Landes region – as well as the Paris region. In some of these areas, the memory of malaria is still very much alive, like in Corsica and Camargue; whereas in others, inhabitants are largely oblivious to the region's malarial past (Mieulet, 2015). The causes behind malaria eradication in these different regions are multifaceted. The draining of wetlands to improve farming and the diking of waterways to improve their navigability and attempt to contain flooding mechanically reduced environments that encouraged the reproduction of mosquitoes responsible for the transmission of malaria. Improvements in health, public sanitation and improved living conditions also reduced the sanitary vulnerability of human populations (see Fouque herein). The spraying of DDT conducted by the American army after the liberation of France was, according to first-hand witnesses, particularly widespread, although it was not systematic. The last cases of indigenous malaria[21] in continental mainland France[22] date back to the late 1940s – and yet widespread mosquito-control policies were only implemented about a decade later. Paradoxically, such largescale mosquito-control policies were initiated at a time when mosquitoes no longer constituted a health risk in these regions.

The implementation of so-called "comfort-based" public mosquito-control policies was initiated in the Languedoc-Roussillon region in the context of largescale tourism development initiatives driven by the "Racine" inter-ministerial mission (MIR)[23] starting in 1963. The

---

21 Regarding the different diseases transmitted by mosquitoes, it is important to distinguish between indigenous cases – i.e. those contracted locally – and imported cases – i.e. those contracted in endemic zones.

22 Meaning strictly mainland France, excluding Corsica which is neither part of mainland France, nor really part of overseas France. Corsica is a unique example. Cases of indigenous malaria were reported until the early 1970s. Mosquito-control measures were conducted by the *Société d'aménagement pour la mise en valeur de la Corse* (SOMIVAC – Planning committee for the enhancement of Corsica), created in 1957, which took over from the DDT treatments done by the American army following the liberation of France. Then, after a short hiatus and alongside the re-emergence of malaria in the 1970s, mosquito control policies were implemented by public authorities.

23 Named after its president, Pierre Racine.

abundance of mosquitoes in the numerous coastal lagoons and wetlands of the region were considered a major obstacle to the creation of large coastal resorts. Comfort-based mosquito-control campaigns – i.e. whose goal was to reduce the nuisance – were awarded to the newfound Entente Interdépartementale de Démoustication Méditerranée (EID-Med), created in 1959 and truly developed starting in 1963. As historian Jean Sagnes (2001) has reported, MIR had direct ties to the prime minister and had complete, unprecedented freedom over its administrative, legal and budgetary initiatives. In the era of great state planning, the debate between economic development and environmental concern took the form of zoning based on territorial specialization. Camargue, which had had a natural reserve since 1927 and became a Regional Natural Park in 1970, was erected as a green belt between the mass tourism of the Languedoc-Roussillon region and the petrochemical industry around the Etang de Berre (Picon, 2008). It was for this reason that it was not subject to mosquito-control prior to 2006.

The early 1960s were an era of technological innovation. The profession of "comfort-based mosquito-control agent" had yet to be invented. To make this happen, different bodies within the public service were solicited, notably military doctors and former colonial administrators. Until the mid-1960s, DDT and then HCH were sprayed on a broad scale, sometimes locally combined with petroleum. These organochlorine treatments mainly targeted adult mosquitoes, whereas petroleum was used against urban larval habitats, since it formed a film on the surface of water that asphyxiated mosquito larvae. One of the first EID-Med managers recalls, "At the beginning, when we were throwing around DDT and HCH, […] I remember when I arrived on the job […] when we went to collect results, and we saw all the flies, dragonflies and all, on the ground, we would say we can't continue like this […] And very quickly, we became aware that things needed to be done differently." With scientific support from the Universities of Montpellier and Marseille, EID-Med gradually focused its mosquito-control strategies on larvicidal treatments using Temephos, an organophosphorus insecticide that is less persistent than organochlorines and which allows for treatments that are more targeted over space and time. Changes in the European regulatory framework later forced EID-Med to abandon the use of Temephos in favour of *Bacillus thuringiensis israelensis*, a biocide described as organic. BTI is less persistent and more selective than organophosphorus products, but is also more expensive and harder to handle.

Other French regions followed in the footsteps of EID-Med. EID-Atlantique and EID-Rhône Alpes were created in 1968 and 1970 respectively. In the Bas-Rhin region, mosquito control came under the authority of a joint association (*Syndicat Mixte*). These mosquito-control operators were regulated by Law #64-1246 of 16 December 1964 on mosquito control. This law gave operators the right to intervene in public and private areas and also stipulated that "owners, renters, franchise holders, farmers and occupants must obey the orders of agents of these departments." The intrusive nature implied in the terms of this law were actually more flexible in the field. From the outset, EID-Med notably used locally-born agents who were extremely familiar with the territory and local population, thus contributing to establishing a detailed cartography of the wetlands where mosquitoes laid their eggs, on the one hand, and facilitating interventions on private property on the other hand.

Until the mid-1990s, the zoning that excluded Camargue from mosquito-control policies was never challenged in an organized or targeted manner (Picon, 2008). Controversy appears to have arisen in the public sphere in 1995 following a change in municipal political power in Camargue (city of Arles). The debate opposed those for and those opposed to mosquito-control, the most active of which were respectively the drivers of the tourism-based economy and local naturalists. The debate was fuelled by a peri-urbanization process in Camargue. City-dwellers indeed began moving to the area in search of a certain quality of life that was challenged by the presence of mosquitoes: e.g., relaxing in their gardens, eating on patios, outdoor entertainment. The debates did not enable a dominant position to emerge, however, and there was a lack of data in the scientific recommendations. Indeed, amidst the modernist euphoria of the 1960s, urban planners had omitted to conduct an environmental assessment prior to implementing mosquito-control. The lack of a "baseline" made it difficult to objectively quantify the impact of mosquito control on ecosystems in the Languedoc and left the field open to partisan positioning.

In 2001, public authorities solicited "public opinion". A sociological survey collected the points of view and ambivalent expectations of inhabitants in Camargue. The presence of mosquitoes was spontaneously mentioned by 54% of inhabitants as the primary inconvenience in the area, much more than geographical isolation, climate, flooding or insecurity. When the question was asked specifically, 64% of those surveyed claimed to be very bothered by mosquitoes. And yet, 67% also claimed to be accustomed to insect bites. Similarly, while 62% of those surveyed described mosquitoes as being a nuisance, 69% felt that the insect played a useful role in nature. Finally, while 66% of those surveyed were unfavourable to mosquito-control across the entire Camargue region, 81% were favourable to mosquito control in inhabited zones. This vernacular taxonomy was constructed on a city/nature dualism that was not biologically unfounded. Among the indigenous species, *Culex pipiens* had indeed long adapted to urban areas, given the water sources rich in organic matter from septic tanks and sanitation networks that provided the biochemical properties sought by females of the species for laying their eggs. The *Aedes caspius* and *Anopheles* species, which lay their eggs in marshes, lagoons and rice fields, are conversely field mosquitoes. The spatialized compromise advocated by local populations could not be implemented, however. These indigenous mosquitoes can indeed easily fly over hundreds of meters or even a few kilometres, thus passing from one type of area to another. The following year, in the context of the European life programme, a sociological survey was conducted in Languedoc-Roussillon (municipality of Aigues-Mortes), in which mosquito control had been conducted for several decades. A large majority of the sample expressed satisfaction with mosquito-control policies and refused to question the way they were conducted. Nevertheless, when asked about the main inconvenience in their living area, the overabundance of tourists was the most-cited factor (47%).

Until the early 2000s, debates over the mosquito-control policies and techniques implemented in mainland France were above all environmental controversies. The terms of these debates opposed proponents of protecting ecosystems and those in favour of human comfort; the authenticity of territories versus their economic development and tourism potential. The health argument periodically resurfaced, with images of children and/or tourists whose bodies were covered in mosquito bites and fevers mentioned by some of the population and certain local doctors. In response, those opposed to mosquito-control criticized an unfair dramatization. They advocated learning to live with mosquitoes, praising traditional Camargue culture incarnated by the emblematic figure of the herdsman, whose traditional attire (boots, long pants, blouse, scarf, hat) was a model to adopt as a stalwart boundary against mosquito bites[24]

---

24 As Bernard Picon (2008) has recalled, herdsmen were historically poor shepherds who lived in barren cabins, wore rags and were highly exposed to malaria. Traditional herdsman attire is a relatively recent invention (early 20th century), inspired by the

(Claeys, 2002; Claeys and Nicolas, 2009). While the few cases of equine West Nile virus identified in the early 2000s in the Rhône Delta were a first sanitary alarm bell, they were not enough to sway opinions. Unlike the United States, the strains circulating in France have not yet resulted in serious or deadly cases in humans.

2005 was a cornerstone year, however, when meteorological conditions encouraged an exceptional proliferation of mosquitoes and in turn spurred lively local reactions. It was also in 2005 that a chikungunya outbreak occurred on Reunion Island, an event widely covered in the national media (Duret *et al.*, 2013, Thiann-Bo Morel and Duret, *infra*). It had also been one year since *Aedes albopictus* – a potential vector of chikungunya and dengue fever – had first been detected in the Alpes Maritimes department. Local naturalists who had always been opposed to mosquito-control in Camargue were not oblivious to this information. They were concerned about the widespread phytosanitary treatments being conducted in Reunion to roll back the chikungunya epidemic (cf. *infra*) and feared that Camargue would be subjected to similar measures in the case of a sanitary crisis. In agreeing to the implementation of localized mosquito-control, naturalists opted for a strategy of controlled anticipation. Mosquito-control as such began in Basse-Camargue in 2006 and was timidly described as "experimental".

For a decade now, this mosquito-control has been renewed annually, under strict scientific monitoring. Swallows, passerines, bats, dragonflies and diptera of the Camargue are the focus of extremely precise longitudinal observation to measure the potential ecological impact of BTI treatments. Sociological monitoring of this mosquito-control initiative has shown the strong local demand for the pursuit and expansion of mosquito-control in Basse-Camargue, as well as acceptance for alternative methods used in addition to rather than as a substitution for the widespread spraying of BTI (Nicolas *et al.*, 2016).

To date, mosquito-control operations in Basse-Camargue have been conducted by EID-Med and are based on longstanding methods used in the Languedoc-Roussillon region. The BTI spraying techniques used in the wetlands and marshes are adapted to the destruction of indigenous mosquito larvae, but they are not effective against *Ae. Albopictus* since its larval habitats are located elsewhere. Unlike the indigenous *Aedes* species

Marquis de Baroncelli who helped codify and institutionalize Camargue folklore as it exists today in local *férias* (fairs).

which prefer ponds and marshlands to lay their eggs, *Aedes albopictus* prefers a profusion of small and large freshwater reservoirs, saucers, pots, vases, fountains, etc. particularly prevalent in urban and peri-urban zones where largescale insecticide spraying cannot be conducted due to technical and legal constraints.

Greening in the South and the re-emergence of sanitary concerns in the North: the permeability and permanence of borders

The decolonization and greening of vector control in the French Antilles

Mosquito control in the French Antilles has always been and remains a health issue. Medicinal solutions have always been favoured when they exist (quinine for malaria, the vaccine for yellow fever). The specificities of arboviruses such as dengue fever have only been taken into account much more recently. For a long time, doctors did not clearly distinguish between arboviruses and other fevers (Darriet, 2014). And, once identified, they were particularly recalcitrant to medical and scientific innovation. The very recent vaccine against dengue fever, with its limited efficiency, has not yet been approved by WHO, and there is still no solution other than symptomatic treatment for chikungunya and Zika virus, for which no vaccines exist.

Policies promoting the spraying of organochlorines and then organophosphorus treatments were initially quite successful in reducing the biomass of *Ae. aegypti* – but only for a short time. In the French Antilles, *Aedes aegypti* indeed proved to be a champion at insecticide resistance (Marcombe *et al.*, 2009). This appears to be one reason for the recrudescence of arboviruses in recent decades (Darriet, 2014). When VC services took stock of the growing inefficiency of organophosphorus treatments in the Antilles, they rolled back their usage, favouring other strategies instead. The main approach chosen and still embraced involves strengthening what was initially called sanitary education. This involves awareness-raising campaigns primarily aimed at the general population and school children. In both Martinique and Guadeloupe, voluntary policies were broadly implemented starting in the 1980s and 1990s. Mosquito and vector control services were particularly active and strived to be original in their communication strategies in accordance with WHO recommendations and pilot projects and/or based on local initiatives. These initiatives took into account the criticism aimed at the normative nature of health education principles and really seized on the notions of "social" and "community mobilization" found in the guidelines and protocols recommended by the large international organizations. Whether official orders or part of the participatory movement of recent decades, these communication methods nevertheless remained devoted to changing the behaviour of the population. And yet the literature has shown that such awareness-raising campaigns do little to nothing towards actually changing people's behaviour (Winch *et al.*, 1992; Moatti and Peretti-Wattel, 2009; Claeys and Mieulet, 2013). The attention given to local populations by contemporary VC services has nevertheless helped to avoid encountering the same hostilities faced by prevention services in the early 20th century. Awareness-raising campaigns have been quite well received and successful in improving the knowledge of populations vis-à-vis vector mosquitoes and arboviruses (Setbon *et al.*, 2008). But the step from being aware to actually implementing measures is still the central stumbling block (Claeys and Mieulet, 2013; Claeys *et al.*, 2016). The fundamental problem encountered in 1911 remains in part unchanged: by opting to target the general population, these VC policies only address a single facet (domestic larval habitats) of the dual-faceted problem already identified in 1908 by the scientific committee directed by Doctor Noc. They continue to almost entirely ignore larval habitats related to the shortcomings of the water conveyance and sanitation networks, as well as urban forms. In 1911 and still today, the decision to focus VC efforts and messages on domestic habitats feels like a last resort effort. The apparent blindness of public authorities discourages the most willing inhabitants and offends the most susceptible who continue to see them as a form of unjustified stigmatization. This sense of stigmatization was exacerbated by the colonial context and racism in the early 20th century. The prevention department at the time was directed by whites for whites. Today, managers and VC fieldworkers are primarily Creoles.[25] The population tends to be open to these West Indian mosquito control operators, while all the while continuing to recover rainwater in barrels that are particularly favourable to the reproduction of *Ae. aegypti* (cf. *infra* Claeys *et al.*) to offset the frequent water shortages from the defective water conveyance network.

[25] Trained nevertheless in mainland France, these people are thus divided between their Creole identity and the influence of mainland France. This identity-based dilemma is not new and was highlighted by psychoanalyst and activist Franck Fanon (1952).

There are also simultaneous calls for better public management, with finger pointing at the lack of maintenance and planning in public spaces (Claeys and Mieulet, 2013). VC services have recently tried to address this urban issue. In Guadeloupe, ARS commissioned a consulting firm to conduct an inventory of urban architectural forms that encourage the formation of larval habitats (ACSES, 2006). In Martinique, the VC department has managed to develop special partnerships with a few municipalities. But these voluntary approaches are forced to contend with active opposition or passive inertia from professionals in the construction and green space sectors, as well as some municipalities.

The urban planning deficit is also as old as the colonization of the French Antilles. The postcolonial Antilles now feel the weight of their poor development: faulty water conveyance and sanitation networks, poor coverage and irregularity in public transportation networks, a prevalence of so-called "spontaneous" or informal settlements. Socio-economic, cultural, ethno-phenotypic, sanitary and environmental inequalities tend to mutually reinforce each other. The prevalence rate during dengue, chikungunya and Zika virus epidemics is a constant reminder of the vulnerability of the poorest and least educated segments of the population (Otmani del Barrio, *infra*). These local populations do not systematically have access to health services when they get sick. But the networks of sentinel physicians implemented by health authorities to quickly detect epidemic outbreaks can only report what they observe in their practices. By the time the alarm bell is sounded, the number of cases is often already high, making it harder to curb the outbreak.

Tourists are less exposed. Most resorts are equipped with air conditioning and hire private operators to regularly conduct mosquito-control operations. The managers of these resorts prefer to act discretely and are loathe to provide their clientele with brochures from the Regional Health Agency detailing the "highly unattractive" epidemic risks.

The introduction of *Aedes albopictus* in mainland France: Ambiguity between comfort-based mosquito-control and vector control

The early 2000s marked the (re-)emergence of health issues in mainland France with the accidental introduction of *Ae. Albopictus*, a potential vector for dengue, chikungunya and Zika virus. *Ae. albopictus* was first identified in 2004 in the town of Menton and quickly expanded its range to cover the entire French Mediterranean coast, followed by the Atlantic coast; it moved up the Rhône Valley into the Alpine lowlands; more recently, it has reached the Paris region and lowlands around Strasbourg (http://www.moustiquetigre.org/).

The rapid expansion of the range of *Ae. albopictus*[26] altered the general sense of comfort that reined along the French Mediterranean coast. From east of the Camargue region to the Italian border, comfort-based mosquito control had always been very localized (Etang de Berre and the Hyères salt marshes, notably), but the low prevalence of wetlands had largely spared the population in this area from such anthropophilic insect attacks. To the west, all the way to the Spanish border, five decades of mosquito control orchestrated by EID-Méditerranée had visibly curbed the nuisance. Yet following the introduction of *Ae. Albopictus*, the population along the entire French Mediterranean coast has expressed growing levels of nuisance. In 2012, 25.94% of the inhabitants surveyed described an increase in the nuisance and 38.47% of respondents claimed they were strongly or very strongly bothered by the insects. Those with homes with a garden and/or patio are significantly more bothered than people who live in apartment complexes (Mieulet 2015). And, along the French Mediterranean coast, single family homes remain socially discriminating: it is primarily the upper and middle classes who lives in homes with gardens. To date, exposure to *Ae. albopictus* bites has had a greater effect on populations that are not socially vulnerable, thus diverging from the dominant scenarios in the French Antilles.

The introduction of *Ae. albopictus* in mainland France quickly resulted in the drafting of a Plan to limit the dissemination of chikungunya, dengue and Zika virus (first drafted in 2006 and regularly updated since). Like in 1908, it was only after these diseases threatened mainland France that the central government truly seized on the issue. To date, indigenous cases of dengue, chikungunya and Zika virus have sporadically been identified and outbreak areas have been controlled (www.invs.santepubliquefrance.fr). Per article L.3113-1 of the code of public health (translated here), dengue, chikungunya and Zika virus require the "mandatory transmission of individual data to health authorities." Following each suspected case of dengue, chikungunya and Zika virus, health authorities mandate a

[26] The scientific literature mentions the potentially aggravating nature of global warming on the proliferation of vector mosquitoes (Gould *et al.*, 2009), while also noting the growing ability of *Ae. albopictus* to adapt to temperate climates (Schaffner *et al.*, 2013).

mosquito control operator to conduct reinforced mosquito-control around the home (and sometimes the main places visited by) the infected individual. These insecticide treatments aim to totally eliminate larvae and adult mosquitoes in order to reduce the risk of an epidemic outbreak, but their effect is very local (neighbourhood scale) and limited in time (a few days) in terms of reducing the nuisance.

The correlation between the introduction of a vector mosquito and the emergence or increase of discomfort tied to the nuisance side of the mosquitoes underscores an ambivalence between health-related and hedonistic claims. In 2012, the nuisance experienced led 74.95% of inhabitants interviewed along the entire French Mediterranean coast to call for the implementation and/or pursuit of mosquito-control by public authorities. Influenced by environmental criticism of insecticides, 47.91% of the sample felt that mosquito control could be dangerous for humans and/or nature. And yet 69.75% of these sceptics were nevertheless in favour of implementing or pursuing mosquito control measures. In 75.12% of cases, the basis of such calls for mosquito control was tied to the health risks related to mosquitoes. And yet only 49.73% of the inhabitants interviewed felt that a chikungunya or dengue epidemic might break out in their department. Moreover, of the 39.77% who felt that there would not be a chikungunya or dengue epidemic in their department, 72.88% were nevertheless in favour of mosquito-control policies.

Calls for mosquito control by the population as such tend to be based on the use of health arguments to justify aspirations for comfort. The use of health-based arguments made it acceptable in eyes of the inhabitants interviewed to sway the balance between environmental concerns and the discomfort caused by *Ae. albopictus*. The boundary between health and comfort is nevertheless quite porous. Indeed, while many indicators converge to underscore the positive influence of the quality of living conditions on health, the scientific literature is confronted with the difficulty of defining and measuring this notion (Le Moigne, 2010). From an epidermal annoyance during an outdoor dinner party to a typical skin infection, all the way to the transmission of dengue, chikungunya or Zika virus – the boundary between comfort and health is porous.

And such taxonomical shifts are further reinforced by a few legal-administrative ambivalences. VC strategies are defined by the government but their implementation falls under the auspices of local authorities who conduct awareness-raising campaigns and mosquito-control treatments. The definition and implementation of strategies to combat nuisance-related problems including so-called comfort-based mosquito control is the sole remit of municipalities, however.[27] And yet the entomological reality does not fit neatly into this French administrative cut-up, since *Ae. albopictus* is both a nuisance and vector.

Could the permeability of the boundaries between comfort and health, as well as the administrative ambiguity between VC and comfort-based mosquito control not ultimately be part of a means of working for the common good: i.e. preventing the dissemination of chikungunya, Zika virus and dengue fever? The limits of such a solution are nevertheless rooted in the notable differences in VC and comfort-based mosquito-control techniques. The use of biocides in the context of VC is limited in time and space. Their goal is to eradicate *Ae. albopictus* around suspected cases during their viremic phase[28] in order to avoid an epidemic outbreak. Such larvicidal and adulticidal treatments reduce the nuisance only over a very short time span and physical area, particularly given that suspected cases of dengue, chikungunya and Zika virus have thus far been relatively rare and geographically scattered in mainland France. Comfort-based mosquito control, on the other hand, involves the spraying of biocides that target the different indigenous species that lay their eggs mainly in wetlands (marshes, lagoons, ponds, etc.). Thus, the different approaches between comfort-based mosquito-control and VC mean that it is not possible to "kill two birds with one stone" and make it technically difficult (if not impossible) to find a common ground in this regard. And yet most of the inhabitants interviewed knew little to nothing about the differences between comfort-based mosquito control and VC. Take for example this inhabitant from a neighbourhood in Nice that had been the target of adulticidal VC treatment who stated, "Once, they came at 6am with a large truck. But they only sprayed the neighbourhood [...] They could have done the whole city. I mean, there's no point, because you can't eradicate all the mosquitoes by only treating a small area."

This type of confusion observed among the population is in part caused by the strategies used in public awareness campaigns. Along the Mediterranean coast in particular, mosquito-control operators and

---

27 Law #64-1246 of 16 December 1964 modified by article 72 of Law #2004-809 of 13 August 2004 on local liberties and responsibilities, article L. 3114-5 of the Code of Public Health.

28 Regarding chikungunya, the viremic phase begins roughly 5 days after being bitten and lasts 7 days on average, whereas the viremic period for dengue begins roughly 2-3 after being bitten and lasts about 7 days (Terrien, 2008).

elected officials tend to mum the word on epidemic risks. The latter are faced with a classic dilemma: convey health prevention information without causing panic or causing secondary effects on the local tourism-based economy. While the local population tends to manipulate health arguments to the benefit of their comfort-based aspirations, local authorities are conversely tempted to manipulate the registers of nuisance and comfort to avoid talking about the health issues related to the proliferation of *Ae. albopictus*. The interviews conducted with decision-makers and operators, as well as during direct observation in feedback meetings and conferences point up different communication priorities between the different actors involved in defining and implementing anti-dissemination strategies. One of the main tensions observed was related to the dilemma between alerting and reassuring. The proximity between local elected officials and mosquito-control operators encouraged the latter to favour reassuring discourse so as not to tarnish the touristic appeal of the region. Public health institutions are on the other hand favourable to the adoption of information campaigns that explain the epidemic risks tied to the introduction of *Ae. albopictus* and the severity of chikungunya, dengue and Zika virus symptoms. This desire to explain the health risks is part of a prevention principle meant to show, among other things, the active involvement of health authorities. This debate between alerting and reassuring has caused some tension between institutional actors. For example, the Occitanie Regional Health Agency has developed its own awareness campaigns alongside those conducted by EID-Méditerranée. Yet recently, the former addressed *Ae. albopictus* as a vector mosquito… whereas the latter addressed it as a nuisance.

While the questionnaire-based surveys conducted in 2010 and 2012 underscore that a majority of the local population is favourable to mosquito-control treatment, some opposition has nevertheless since emerged locally. This has mostly focused on VC intervention surrounding suspected cases of dengue and chikungunya. One of the most recent examples involved the mobilization of local inhabitants who physically blocked the passage of spray vehicles loaded with insecticides. This occurred in a municipality in the Rhône-Alpes region that has been part of a "zero pesticides" initiative in public green areas for several years. The double discourse of a polycephalous public authority advocating the removal of pesticides in gardening on the one hand and initiating intense insecticidal treatments in the name of vector control on the other hand is a prime example of how incomprehension/reactions can be spurred in some parts of the population.

Conclusion: the unfindable "community of destiny"[29] between mainland and overseas France

As Shaw, Robbins and Jones (2010) have argued, "mosquito management strategies are highly differentiated, entwined not only with history, technology, and available chemical resources but the spatial ontologies that enable their mobilization." The recent history of mosquito-control policies in mainland and overseas France – whether comfort-based or for vector control purposes – reveal the inclusion of environmental concerns. Over the span of a just a few decades, the large-scale spraying of powerful and non-selective insecticides has given way to more targeted operational approaches and more selective biocides. The drivers of such change are ideological (increased ecological consciousness), legal (European context), as well as economic (the cost of treatments) and bio-technical (the resistance of mosquitoes to insecticides and the colonization of anthropized areas). The emergence of vector-borne diseases in mainland France and their increase overseas have however begun to counter the greening trend in mosquito-control policies. In the face of suspected cases in mainland France and the multiplication and diversification of epidemic outbreaks in the French Antilles, a chemical arsenal is regularly deployed in the name of sanitary emergency, thus trumping ecological concerns. And yet, at the same time, the capacity of *Ae. albopictus* and *Ae. aegypti* to adapt and resist have swayed the anthropocentric "obvious answer" from the absolute need to address a sanitary emergency to the importance of protecting nature. Indeed, an entirely insecticide-based approach in the name of vector control is not only potentially dangerous for human health and for biodiversity, but it is also increasingly inefficient.

This comparative socio-historical examination of mosquito control on both sides of the Atlantic indeed reveals a natural environment that remains natural in its ability to no longer adapt independently from mankind (the classic opposition of natural/artificial), but in response to anthropic actions. Mosquitoes, their resistance, colonization and proliferation eloquently express the ability of nature to change, "its autonomous potentiality for evolution, change, mutation, future activity and creativity," as Céline Granjou (2016, translated here) has written. In this respect, the formula espoused by Prochiantz and Descola (2014,

29 The expression "community of destiny" was coined by Jonas (1990).

translated here) is a concrete biological reality, "There is a human history of animals in the same way there is an animal history of humans."

In the face of increased nuisance and the emergence of vector-borne disease on the French Mediterranean coast, public authorities reacted quickly. Entomological monitoring, health monitoring and national plans were quickly rolled out, accompanied by their share of ministerial visits and media coverage. In the face of invading miasmas from the South, mainland France has invested and taken measures to protect itself. In the French Antilles, on the other hand, vector mosquitoes have long been a part of everyday life for local authorities and the population. At the height of the worst epidemics, the French Antilles were obviously the focus of a few compassionate ministerial trips and on the receiving end of "increased sanitary efforts," but the underlying causes were never addressed: the under-development or more precisely the poor-development of territories marked by social and spatial inequalities exacerbated by the post/neo-colonial context.

The discomfort of the rich in the North and a fatality for the poor in the South? *Ae. albopictus* and *Ae. aegypti* are now present on both sides of the Atlantic: the former in mainland France and the latter in the Antilles. Today, all in-bound flights from a destination in the Antilles undergo insecticidal treatment to reduce the risk of introducing tropical pathogens and vectors to mainland France. The same precautions are not taken with aircraft landing in the Antilles that could unwittingly be importing a few *Ae. albopictus* mosquitoes from mainland France. Given the limited efficiency of such treatments this example is likely anecdotal, but the symbolism nevertheless remains: everything is done as though health protection measures are taken to protect a sanitized North from the diseases of the South, while protecting the South from incursions from the North remains inconceivable. This incapacity to think reciprocally about endangering the other is an extension of our inability to conceive of our community of destiny – between mainland and overseas France, between the North and South, between whites and blacks.

References

Abenon L.R. (1992). *Petite histoire de la Guadeloupe*. Paris, L'Harmattan.

Brissac D. (2011). « L'écologie urbaine dans les îles de la Caraïbe : le cas de la Guadeloupe. La dynamique des écosystèmes produits par les espaces urbains, la ville et les quartiers en étalement urbain », In P. Teisserenc, R. Etien et P.-Y. Chicot (dir.). *La recomposition territoriale : un défi pour la Guadeloupe* (205-218). Pointe-à-Pitre, Presses de l'Université des Antilles et de la Guyane ; Paris, Publibook.

Burner M. (2015). « L'occupation à risque du territoire, traduction contemporaine d'un contexte géo-climatique et historique », In J.-C. Huc et M. Etna (dir.). *Éclats de temps. Anthologie des événements climatiques extrêmes de la Guadeloupe* (332-343). Le Gosier (Guadeloupe), PLB Éditions.

Buttel P. (2007). *Histoire des Antilles françaises*. Paris, Perrin, 423 p.

Chan K.L., Ho B.C. and Chan Y.C. (1971). "Aedes aegypti (L.) and Aedes albopictus (Skuse) in SingaporeCity", *Bulletin de l'Organisation mondiale de la santé*, 44(5), 629-633.

Claeys C., Azibi L., Reine A., Bayle S., Dufour A., Mieulet E., Perroni M. and Spinnato J. (2009). « Moustiques et démoustication dans les Alpes Maritimes : analyse sociologique des discours et pratiques des habitants. Première phase : Enquête qualitative exploratoire et préparatoire. Rapport de recherche », *E.I.D.-Méditerranée*, Responsable scientifique : C. Claeys.

Claeys C. and Mieulet E. (2012). « Suivi sociologique de la démoustication expérimentale de Salin-de-Giaud et Port-Saint-Louis-du-Rhône, résultats 2012, bilan 2007-2011, Contrat de recherche *Parc Naturel Régional de Camargue et Conseil Général des Bouches-du-Rhône* », Responsable scientifique volet sociologique : C. Claeys, www.parc-camargue.fr.

Claeys C. (dir.) (2016). « Prolifération d'espèces littorales terrestres et marines à fort enjeux environnementaux et sanitaires : une comparaison métropole (côte méditerranéenne) et Outre-mer (côtes antillaises) (PROLITENSAN) », *APR 2012 Quels littoraux pour demain ?*, Fondation de France, Rapport final.

Claeys C. and Mieulet E. (2013). "The spread of Asian tiger mosquitoes and related health risks along the French Riviera: An analysis of reactions and concerns amongst the local population", *International Review of Social Research*, 2(3), 151-173.

Claeys C. and Mieulet E. (2013). Volet sociologique du *Programme Européen Life + 08/ENV/F/000488*, coordonné par l'EID-Méditerranée, « Lutte contre les moustiques nuisant et vecteurs de maladies : Proposition d'une gestion intégrée compatible avec le développement durable IMCM (Integrated Mosquitoes Control Management) » (2010-2013).

CLAEYS-MEKDADE C. (2000). *Les conflits d'aménagement : rapports à la « nature » et rapports sociaux. La Camargue, un cas révélateur.* Thèse de doctorat en sociologie, Université de Provence.

CLAEYS-MEKDADE C. (2002). « Les controverses relatives à la démoustication de la Camargue : rapports à l'animal et au territoire », *Espaces et Sociétés*, 110-111(3), 147-166.

CLAEYS-MEKDADE C. and MORALES A. (2002). « Moustiques et démoustication : une enquête sociologique auprès des Arlésiens et des Camarguais », *Rapport final sur l'étude d'impact d'un éventuel traitement au BTI sur le territoire du Parc Naturel Régional de Camargue*, DESMID-IMEP, contrat PNRC, responsable scientifique : B. Picon, 6-72.

CLAEYS-MEKDADE C., CORSAND, L., NICOLAS L. and SCHLEYER-LINDENMANN, A. (2002). « Être ou ne pas être entre les deux bras du Rhône : identité(s) camarguaise(s) aujourd'hui », *Faire Savoir, Revues, Sciences humaines et sociales en région PACA*, 2, 25-32.

CLAEYS-MEKDADE C. and NICOLAS L. (2009). « Le moustique fauteur de troubles », *Ethnologie française*, 39(1), 109-116.

CLAEYS-MEKDADE C. and NICOLAS L. (2002). « Être ou ne pas être démoustiqué, enquêtes ethnologique et sociologique comparatives : Camargue et Petite Camargue », *Contrat DESMID-EID, Programme européen Life-environnement N°life 99 env/F/000489*, Responsable scientifique : B. Picon, 104 p.

DARRIET F. (2014), *Des moustiques et des hommes : chronique d'une pullulation annoncée*, Paris, IRD.

DURET P., CUBIZOLLES S. and THIANN-BO MOREL M. (2013). « La crise sanitaire du chikungunya : une épreuve de recomposition des rapports sociaux à La Réunion », *Sociologie*, 4, 3.

HANSON S. and CRAIG G.J. (1994). "Cold acclimation, diapause, and geographic origin affect cold hardiness in eggs of Aedes albopictus (Diptera: Culicidae)", *Journal of Medical Entomology*, 31(2), 192-201.

JONAS H. (1990). *Le principe responsabilité.* Paris, Cerf.

ICHECK V. (2004). *La lutte contre les moustiques à la Martinique: approche de l'évolution institutionnelle et technique du XVIIe à nos jours.* Mémoire de Master d'Histoire, Université des Antilles.

LE MOIGNE P. (2010). « La qualité de vie : une notion utile aux sciences sociales ? », *Sciences sociales et santé*, 28(3), 75-84.

Marcombe S., Poupardin R., Darriet F., Reynaud S., Bonnet J., Strode C., Brengues C., Yébakima A., Ranson H., Corbel V. and David J.P. (2009). "Exploring the molecular basis of insecticide resistance in the dengue vector Aedes aegypti: A case study in Martinique Island (French West Indies)", *BMC Genomics*, 10(1), 494.

Massé R. (2007). « Les sciences sociales au défi de la santé publique », *Sciences sociales et santé*, 25(1), 5-23.

Mieulet E. (2015). *La prolifération de moustiques vecteurs sur le littoral méditerranéen et dans les départements d'Amérique: enjeux environnementaux et sanitaires*. Thèse de Sociologie, Aix-Marseille Université, 425 p.

Mieulet E. and Claeys C. (2015). « Transferts de responsabilité entre sphère publique et privée : le cas de la prévention des épidémies de dengue en Martinique et en Guyane », In A. Meidani, É. Legrand et B. Jacques (dir.), *La santé : du public à l'intime* (87-103). Rennes, Presses de l'EHESP.

Moatti J.-P. and Peretti-Wattel P. (2009). *Le principe de prévention. Le culte de la santé et ses dérives*, Paris, La Martinière/Seuil, 112 p.

Sagnes J. (2001). « L'aménagement touristique de la côté du Golfe du Lion », In J. Sagnes (dir.), *Deux siècles de tourisme en France* (27-53). Perpignan, Presses universitaires de Perpignan.

Sainton J.-P. (2015). *Histoire et civilisation de la Caraïbe*. Tomes 1 et 2. Paris, Karthala.

Salvan M. and Mouchet J. (1994). « Aedes albopictus et Aedes aegypti à l'île de La Réunion », *Ann. Soc. belge méd. trop.*, 74, 323-326.

Schaffner F., Medlock J.M. and Van Bortel W. (2013). "Public health significance of invasive mosquitoes in Europe", *Clin Microbiol Infect*, 19(8), 685-692.

Vergès F. (2006) « L'Outre-mer, une survivance de l'utopie coloniale républicaine ? », In P. Blanchard, N. Bancel et S. Lemaire (dir.), *La fracture coloniale. La société française au prisme de l'héritage colonial* (69-76). Paris, La Découverte.

Winch P., Kendall C. and Gubler D. (1992). "Effectiveness of community participation in vector-borne disease control", *Health Policy and Planning*, 7(4), 342-351.

The chikungunya outbreak in Reunion: epidemic or environmental crisis?

Marie Thiann-Bo Morel

UMR 228 Espace-Dev, Université de La Réunion

Introduction

The first cases of chikungunya were reported in Reunion Island, a French *département* located in the Indian Ocean, in 2005.[30] On the island, the main vector is *Aedes albopictus* or the "tiger mosquito". In 2007, there were 267 deaths in Reunion due to chikungunya (Brucker *et al.*, 2007). Yet two years earlier, public health managers had been slow to realize the extent of the epidemic. At the time, little was known about the disease. Nobody knew that it could be transmitted from mother to child (Taglioni, 2006, 2011) or clearly understood the many complications it causes, which vary between patients but would permanently handicap the island's economy. As the number of cases soared in late 2005, public health authorities were taken to task for their inaction. And as the epidemic reached its peak in March 2006, this locally unprecedented health crisis attracted massive media coverage.

Rather than producing another public health analysis, this chapter returns to the epidemic from a new perspective and applies an environmental framework to the understanding of events. We believe that the chikungunya crisis was environmental from the outset, insofar as the health risk was constructed as an environmental wrong which

[30] Since the first reports of chikungunya in 2005, Reunion experienced an "attack rate of 35% of its population or 266,000 cases, including 876 serious cases, 224 paediatric cases, 44 neonatal forms, 222 severe forms in adults with an 11% death rate, totalling an estimated excess mortality of 267 deaths for the island" (Brucker *et al.*, 2007).

local populations seized upon, claiming an accumulation of unequal treatment. The chapter follows the expression of feelings of injustice to understand how this environmental crisis was forged.

How did the interest in the environment emerge during the health crisis? First, by the public authorities' proposed solution – mosquito eradication – being turned into a problem. We will begin by examining how mosquito eradication was gradually constructed as an additional risk. We will look at how the mosquito eradication plans were denounced and their organization brought under attack. This denouncement expanded into a fierce controversy over the toxicity of the products used. Warnings against widespread chemical use led the population and local authorities to refuse mosquito control treatments.

Next, we will see that the environmental question was also brought to bear in justifying alternatives to the mosquito control plans: "healthier" solutions were proposed to satisfy the demand for "biological control". We will try to situate these professedly more "ethical" solutions within the power relations at work in a postcolonial society. For in Reunionese society, the spectre of the economy prevented talk of contagion and evoked the scars of the colonial period.

1. Literature review and conceptualization

1.1 Reunion Island and the difficulties of describing a "postcolonial" territory

Reunion Island is a French island located in the Indian Ocean, 11,000 kilometres from Paris. Formerly a plantation colony, Reunion became a *département* in 1948.[31] This "gentle" transition meant the application of French republican principles (including the equality of all citizens) and the launch of "catch-up" plans for the island's development. Even so, major social disparities persist in Reunion (high levels of illiteracy, unemployment, etc.). The island was populated by successive waves of immigration. The

[31] France is divided into administrative subdivisions: regions and *départements*. The *département* is an administrative division governed by a departmental council (called the *conseil général* in French) with an elected president. The regions each include several *départements* and are governed by a regional council (*conseil régional*) with an elected president. Reunion is a peculiar subdivision as it has dual administrative status: it is both a *département* and a region, under the authority of the prefect and of the presidents of both the regional and departmental councils.

"indigenous" population is now made up of the descendants of slaves and "*engagés*", or indentured workers (mostly from the Indian Ocean region), and people from Metropolitan France, known locally as "*Zoreils*."[32] Reunionese society can be considered polysegmental in the Maussian sense of the term (Ghasarian, 2002; Thiann-Bo, 2009). As such, ethnic and phenotypical considerations help shape the understanding of social phenomenon for individuals socialized within it. Social relations are made complex by the ethnic mixing of populations (and the multiple affiliations that this generates) as well as by ambiguous relations with France, which is still known as the "Metropole". The history of Reunion written from the French perspective has long glossed over this complexity, masking it with a naïve melting-pot view of these populations from different backgrounds. Here, to grasp the complexity of Reunionese society, we will draw on the theoretical framework of postcolonialism.

Figure 1. *World map and Reunion.*

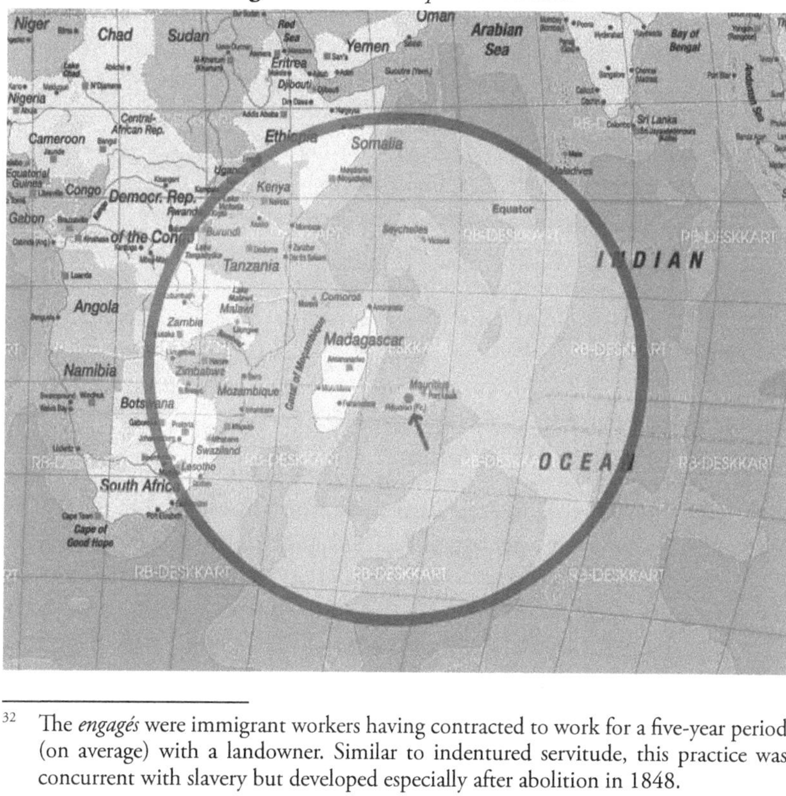

32 The *engagés* were immigrant workers having contracted to work for a five-year period (on average) with a landowner. Similar to indentured servitude, this practice was concurrent with slavery but developed especially after abolition in 1848.

The origins of post-colonial studies can be traced back to two fields: literary criticism and literature (Collignon, 2007). The discipline seeks to situate post-colonial societies not only temporally, after colonization, but *beyond* colonization, that is, by extricating the study of these societies from the dominant Western perspective. Having identified the founding paradigms of the colonial period, such research envisages different ways of going beyond this interpretative framework. For example, it examines how actions revolve around the "centre/periphery pair as the key to reading the world" (Collignon, 2007, 4), with the Reunionese "us" opposing the "them" of the continent (decision-makers being referred to as "*Zoreils*"). We will pay particular attention to the designations of "Others" and analyse the references to indigenousness which certain actors used to their advantage. According to Retière (2003), indigenousness is used to name the symbolic resource of being a native of the country in a competitive society. Focusing on references to indigenousness, *Retière argues*, helps situate readings of the world and definitions of "legitimate actors" (Retière, 2003). Ultimately, this means considering where the actors are talking from and how the voices of the Reunionese are situated in *their* worlds. Reunion, in its relationship with "the rest" of the Indian Ocean, constitutes a western enclave situated between Africa and Asia. Compared to Mayotte, another French *département* in the Indian Ocean, Reunion is the "Monaco of the Indian Ocean" (Gauzères, 2006). Mayotte is an exception in Overseas France. It is the only French overseas *département* not to have a benefits system aligned with mainland France, which constitutes a breach of the principle of equality enshrined in the French social model (Marie, 2014).

1.2 Reunion during the chikungunya crisis

The academic literature on this health crisis is abundant and highlights how it was framed as a mosquito problem (Dupé, 2015; Metzger, 2009; Watin, Metzger, Taglioni and Idelson, 2009; Taglioni and Dehecq, 2009; Setbon and Raude, 2008; Watin, 2007). Moreover, the main mistake of which public health authorities are accused is to have seen the first reports of chikungunya (in 2005) only as a mosquito problem. The determination to wage war on mosquitoes can thus be seen as a paradoxical obstinacy worthy of those Paul Watzlawick condemned for demanding "always more of the same". In times of crisis, we do "more" of what we do best in "normal" times. The literature points out that the comparisons with well-known epidemics of dengue (the first cases in Reunion were reported in 2004) formed a cognitive straitjacket (Watin, Metzger, Taglioni and Idelson, 2009), preventing authorities from predicting and preparing for the disaster. The same literature analyses how vector control (VC), considered a tried and tested method, was reactivated as a lethal weapon that transformed the solution into a problem, despite persistent warnings. Indeed, framing the crisis as a mosquito problem relegated the management of people (both the sick and those not yet affected by the disease) to the margins of the health issue.

In comparison with Mayotte, the press judged that Reunion failed to realize the extent of the danger. The infection rate in Mayotte was similar to that of Reunion: "In fact about 35% of the population, a similar rate to Reunion and Mauritius, has been affected by chikungunya since the beginning of the epidemic, i.e., about 40,000 to 45,000 people and not the 7,148 cases officially reported" (Watin *et al.*, 2009, 45). Yet while Mayotte was deemed to have successfully contained the disaster by initiating a community response, the Reunionese authorities were accused of incompetence in their management practices and seemingly struggled to look beyond familiar solutions. The press emphasized what has since been referred to as the authorities' "structural and cyclical vulnerabilities" (Watin *et al.*, 2009, 51-53) and depicted a territory caught in the throes of its modernity. The economic development of an island seen as a post-colonial society masked persistent social inequalities and prevented the "Monaco of the Indian Ocean" from anticipating the consequences of an epidemic which struck the most fragile strata of society first of all.

Figure 2. *Historical timeline.*

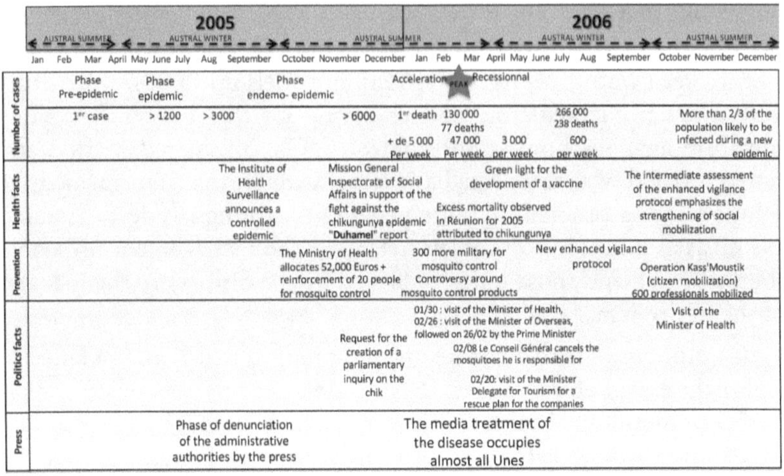

1.3 Understanding social mobilization on environmental issues

In this chapter, we will see how mosquito eradication, proposed as a solution to prevent and reduce the crisis, was paradoxically constructed as an additional risk during the epidemic by the Reunionese population. According to Francis Chateauraynaud (2008), the notion of uncertainty is central to understanding the construction of risk. Chateauraynaud holds that uncertainty has three registers, themselves functions of three different "levels". On the ontological (fundamental) level is the register of indeterminacy. On the cognitive level, the calculations required for "risk taking" relate to the register of indecision. Finally, on the level of values, risk hinges on a register of anxiety (Chateauraynaud, 2008). Uncertainty involves a state of anxiety that compels individuals to be vigilant. Here, we will see how the actors' vigilance enabled them to call mosquito eradication programmes into question based on the potential danger for the environment.

To understand the social mobilization concerning mosquito eradication, we will also use the theoretical framework of environmental justice. This framework focuses on the relationship that populations form with the environment when they express feelings of injustice (Charles *et al.*, 2007). Like Valérie Deldrève (2015), we will consider environmental inequalities as social inequalities; that is, as experiences of environmental wrongs entrenched in relations of social subordination. We will see how, in the post-colonial society of Reunion, the social movement that emerged around the critique of mosquito eradication programmes made "the environment an issue not of conservation, but of social justice" (Lejeune and Drique, 2017a, 111). This theoretical doctrine takes as its "empirical basis the poor communities of colour (who) suffer disproportionately from the consequences of environmental degradation" (Lejeune and Drique, 2017b, 15). Thus, this chapter examines the sense of justice of the Reunionese populations who denounced an *unjust* environmental disaster.

2. Methodology

The press was a key player in the chikungunya crisis (Watin, 2007). "Beyond the medical and health implications, the chikungunya crisis resulted as much from the spatial spread of the virus in the population as from the institutional and administrative management of the phenomenon, by the successive incorporation of new actors and the uncontrolled media coverage of the event" (Watin, 2008). We focus on the peak epidemic period and have defined the temporal scope of our investigation accordingly.

We selected three print newspapers: *Le Journal de l'Ile (Jir.), Le Quotidien (Quot.)* and *Témoignages* (*Tem.*). These three newspapers had the widest circulation at the time of the epidemic; digital news media did not yet have the scale and scope it does today.[33] To a certain extent, these papers were a direct echo of popular views (this is relative, because content has been put into writing and selected according to the editorial line of the newspaper in question).[34] The *Jir.* and *Quot.*, for example, each established a special "chik" letters section so readers could express themselves on the topic.

[33] "In 2011, homes were twice as likely to have an Internet connection than in 2006 and the cost of Internet subscriptions accounted for an increasing share of fixed expenses [...] Internet access options have increased since 2006 with the development of broadband, smartphones and 3G. Reunionese households have equipped themselves [...] The rate of Internet subscription among households in the lowest income quintile is now 28%, compared to only 3% in 2006." Source: Michel Brasset, 2014, Insee flash Réunion published on 25/08/2014.

[34] To get an idea of *popular* thought, it is worth listening to Radio FreeDom's archives, which dedicate a good deal of airtime to the free expression of listeners.

Témoignages holds a special place in this corpus. As a partisan newspaper, it focused mainly on the parliamentary activity of elected members of the Communist Party of Reunion, and provides insight into how the voice of islanders directed towards national authorities did or did not illustrate our hypothesis. The following table summarizes the corpus in terms of the number of pages.

The corpus is referenced as follows: first letters of the newspaper (*Tem.*, *Jir.* or *Quot.*) followed by the date (DD/MM/YY).

Table 1. The press corpus in number of pages

NEWSPAPER	JAN 2006	FEB 2006	MAR 2006
Le Journal de l'Ile (*Jir.*)	97	292	72
Le Quotidien (*Quot.*)	57	283	93
Témoignages (*Tem.*)	95	146	46

We read this regional daily press (Watin, 2008) in its entirety over three months (from 1 January to 31 March 2006) to capture what was covered in the press at the time.[35] The aim of this was to contextualize the chikungunya epidemic in relation to the news reported during the period and compare its media treatment to other events or centres of interest that marked the period. Reading the papers in full ensured that none of these contextual effects were missed. This perspective offsets the accusations of alarmism and sensationalism (Gaüzère, 2011) frequently levelled at the press.

This method of data collection was possible because it covers a short period – the period surrounding the peak of the epidemic – and focuses on its first key developments – the decision to conduct widespread mosquito control and emergency visits by health ministers, the chemicals used and the deaths directly or indirectly attributed to chikungunya. The data time-frame is defined by the "epidemic peak" (the first death formally and directly attributed to chikungunya was announced in February 2006; the epidemic peaked in February 2006) (ANSES, 2013), the period considered after the event to be the worst phase of the crisis.

35 Excluding racing pages and classified ads.

3. Findings

Our analysis shows that over the period studied, the environment was no marginal concern. Its role was never minimized in the media. On the contrary, the environment always featured alongside health and social issues as a factor of prime importance in handling the crisis.

The construction of an environmental wrong originates in a categorical rejection of the administration's solution to the crisis. Two conditions made this rejection possible: the lack of coordination between authorities for mosquito eradication and the controversy over the toxicity of mosquito control products. These parallel movements help to understand how and why a concern for the environment emerged. Let us first look at the suspicion that the massive mosquito control operation was poorly organized, which contributed to turning the solution into a problem.

3.1 How the solution became a problem: the making of the mosquito control risk

In late January 2006, almost a year after the first disturbing reports, the massive mosquito control campaign began – massive due to its scope and the number of actors involved. On 19 January, the prefect announced the new control plan. Mosquito eradication was to start with the five cities most affected by the epidemic (Saint-Louis, Saint-Denis, Saint-Pierre, Sainte-Rose and Sainte-Anne). A ministerial taskforce report written at the very start of 2006 to inform the response to the epidemic (the "Duhamel" report, 2006) highlighted the need for coordination between actors, but this sound advice had little effect. To facilitate the "massive" scale of the mosquito control operations, the prefect brought in the army.[36] Twenty soldiers accompanied by two agents from the regional health and social care board, or DRASS (for *Direction régionale des affaires sanitaires et sociales*), worked at night to treat areas according to the number of reported cases, the worst-hit being treated first (*Jir.* 21.01). The proliferation of actors reassigned to mosquito control had a

[36] The prefect is a senior government official whose mission is to implement government policy. Prefects are appointed by decree of the President of the Republic and represent the state in their administrative division, the *département* or region. As an appointed position, the office of prefect conflicts with the elected positions of the regional and departmental presidents.

detrimental effect. The prefect wanted a community response but did not create the conditions for it and nonetheless expected each local authority to participate in the war effort. Lacking the necessary workforce in-house, the local authorities hired farmers and workers under "green job" and government work schemes to increase the numbers of potential mosquito control personnel. It was this overzealousness that triggered the crisis. Local authorities proclaimed emphatically in the press that they had not been consulted. Some town councils were forced to let the army through before they had had time to warn their constituents (*Quot.* 11.02, p. 15).

This was the case for the mayor of Saint Joseph, a municipality in the south of the island. For this mayor, mosquito eradication was "forced". He felt he had been made "an accomplice to a policy whose effects are not understood by the population [...] if someone sleeps with a window open, to whom will he or she turn? I don't want to be a guinea pig mayor, nor one to stand by and let my constituents pay the price for a hasty policy." (*Quot.* 11.02, p. 15).

The mayor of Saint Joseph balked at this lack of information and refused to allow mosquito control. Journalists marked him out as engaging in "resistance" and others soon joined him in refusing the mosquito control operations initiated by the prefect.

The deployment of the army frightened people – due to its scale on the one hand, but also to the impossibility of citizen opposition. A number of recriminations were reported in the press. On 13.02, for instance, *Jir.* reported that two women had been fined 3,700 euros for refusing the teams of insecticide sprayers access to their garden. "I have children and animals, I feared for their safety," they said (*Jir.* 16.02, p. 15).

The citizens concerned cited this right to oppose mosquito control, believing that the methods were not transparent and the precautionary principle must therefore be applied. The argument of protecting vulnerable people does evince a concern for environmental health. But it also shows that the population constructed the mosquito control risk through the register of anxiety (Chateauraynaud, 2008) and, in doing so, raised the question of whether the cure for the epidemic was worth putting human health at risk.

> We have a product that is potentially unsafe, is not proven to be effective, and we are supposed to keep on using it as if nothing were amiss? Well as far as I'm concerned, you do not play with children's health, especially when the cure is worse than the disease. (*The mayor of Saint Paul* quoted in JIR, 08.02)

Under the circumstances, the mayor decided to stop treating buildings and schools with Fenitrothion immediately. There are two possible interpretations for these reactions. On the one hand, the "rebels" could be seen as engaging in civil disobedience. Citizens may knowingly make an illegal decision in the name of a higher principle. But these acts can be interpreted in another way: as Norbert Elias (1985) argued, a high degree of self-restraint in accepting the law is acceptable only if accompanied by a significant sense of decision-making freedom. A lack of freedom of choice leads to refusal.

There was another reason for some town councils' suspension of massive mosquito control: they had already undertaken certain measures within their competence in environmental matters to combat chikungunya.[37]

The fact that these operations coincided with the end of the school holidays brought the disorder to a head and created an unfavourable situation for the authorities.[38] How could the organizational constraints of a new school year be reconciled with those of mosquito control? With the school staff away, who would treat school grounds (especially when they were not on the list of areas to be treated)? How would people be trained, given the limited time available (one week)? How long before the children arrived should the treatment be applied? The new term was postponed for a week to allow time for schools to be treated, during which time the National Education trade unions (called "TOSs") called for a strike. This made matters even worse for the prefect, who was in charge of coordinating the mosquito control plans.

The press added to the uproar by accusing local authorities of populism. Under the guise of helping spray mosquito control products, some town councils hired locals hand over fist without taking the time to train them. The press questioned the competence of these employees (often farmers or young people on government job schemes called "*contrats d'avenir*") and criticized municipalities for their opportunism. Mosquito control was seen as an opportunity to provide jobs and satisfy voters. The councils countered this accusation of populism by citing the lack of information from the prefecture: training sprayers? "We were told it would only take a few hours!"

---

[37] Information from this period indicates that three councils suspended mosquito control: Le Port, Saint-Paul and Saint-Joseph.

[38] In the southern hemisphere, the summer holidays are from mid-December until the end of January.

The actors all blamed each other to avoid responsibility for the poorly coordinated mosquito control campaign. This buck-passing took place against a backdrop of controversy over the future of preventive health services, which had for several years volleyed back and forth between the state and the *département*. The *département* was afraid that a decentralized preventive health service would be reinstated without the necessary resources.

> In Reunion, until 1986, mosquito control was regularly undertaken by local DRASS services under the authority of the departmental council. DRASS was then nationalized and came under the control of the state. The result? Such a dearth of resources and manpower that the fight against mosquitoes was abandoned for lack of combatants. (*Tem.* 07.03)

This lack of funding for the DRASS preventive health service was reported by the ministerial delegation (Duhamel report) and analysed by Sandra Dupé in her thesis. Dupé showed how the *département* became incapable of effectively dealing with mosquitoes (Dupé, 2015). One year before the 2006 chikungunya epidemic, the president of the departmental council was forced to resume responsibility for the preventive health service. In her press releases, she quoted a letter to the Minister of Health requesting postponement of this decentralization, which was refused.

> In view of the plan to transfer this competence (preventive health) to the departmental council, the local authority officially informed the Minister of Health by letter dated 24 May 2005 that 'The prefect of the Region and the Department wishes to retain full responsibility for mosquito control within his services, including the field operations currently carried out by the DRASS preventive health service'. (*Quot.* January 2006)

Trust in both local and state government was doubly eroded. First, the lack of qualifications of the "professionals of convenience" acting as sprayers was regularly denounced in the press. Second, certain municipalities' refusal of large-scale mosquito control operations was enough to discredit the chemicals chosen by the authorities once and for all, especially as large variations in dosage were reported in the press:

> When they tell us one litre, it's better to use a litre and a half isn't it? Who wouldn't? (Quote from a farmer involved in mosquito control, *Quot.* 10/2/6, p. 11).

So who *could* spray the pesticides? *Témoignages* wrote that the use instructions for Fenitrothion stipulated 216 hours of training: "Is anyone aware that in Reunion, the people spraying this product do not have the qualifications required to protect others and protect themselves?" (*Tem*, 4.01) Even the competence of the army, which until then had coasted on the reputation of military rigour, was called into question. "Spraying insecticide is a job for professionals" (*Quot.* 08.02). The question of correct dosage was also raised. Who measured the dose and mixed the insecticide? Did the sprayers receive the recommended supervision? According to experts consulted by the press, sprayers had much better protection than the average person using an insecticide spray at home. Moreover, while the unions were discussing a stoppage, school employees raised doubts over whether their protection was being considered, and many refused to spray their schools on the grounds that they were not adequately trained or properly protected. Why did some have a hazmat suit while others were expected to spray with their bare hands and no mask?

The Duhamel report had raised this issue. It recommended that special attention be paid to the sprayers' training-training which, already in early 2006, "needed improvement".

The mosquito control campaign was a disappointment. The media blamed the lack of coordination between actors exacerbated by a multi-layered bureaucracy. In the urgency of the situation, the decentralized state services traditionally in charge of mosquito control, the army, mediators, and municipal agents became entangled in a maelstrom widely relayed by the media. Who was doing what? Who organized and who carried out? Who had authority? Who was informing? Who was trained for the purpose? The media implied that the organizational failure was so absurd it must have been deliberate. The incompetence of the eradication teams must have been planned. The coordination strongly recommended by the Duhamel Report did not happen, making the incompetence seem all the more obviously intentional since it had already been pinpointed. This context of suspicion vis-à-vis the administration set the stage for the press and the public to question the crisis management methods used. It thus contributed to shaping mosquito control as a risk. To use the semantics developed by Olivier Borraz (2009), the shaping of this risk originated in a breakdown of familiarity. It seems to be based on several existing beliefs.

The all-too-easy comparison with familiar dengue outbreaks (including one in 2004) (Watin, Taglioni, Idelson and Metzger, 2009) made it impossible to think of the unexpected. According to Pascale Metzger, "The problem posed by chikungunya was immediately likened to the experience of the dengue epidemic that occurred in Reunion Island in 2004" (Metzger, 2009). She suggests that the delayed response to chikungunya was due to the powerful cognitive effect produced by the analogy with the dengue epidemic (Taglioni, Watin, Metzger and Idelson, 2009) and implicates the establishment's beliefs. Indeed, from the outset, the Reunionese health authorities classified chikungunya as an environmental problem that therefore required an environmental solution. Yet the press pointed out early on that the chikungunya crisis was not just a mosquito problem: "The presence of mosquitoes is not enough to raise fears of an epidemic in mainland France" (*Jir.* 13.02.06, p. 7). The controversy over state incompetence caused a rejection of the system of meaning advocated by health authorities thus far and the refusal of mosquito eradication. Associations were subsequently formed that approached the problem differently, particularly from the angle of environmental wrong.

Furthermore, the pre-epidemic period seems to have been governed by the belief that Reunion was a developed society, safe from the public health tribulations affecting the rest of the tropical region. The memory of deadly malaria epidemics in Reunion had faded, and there was a strong belief in the islanders' relative good health and in community solidarity. These beliefs assumed that the health system of a developed country provided protection. They prevented the structural and cyclical vulnerabilities of health management bodies from being identified and addressed (Watin, Taglioni, Metzger and Idelson, 2009). But it soon became clear that the solutions proposed were unsuited to a tropical island environment. Reunion sank into an uncertain period, where health authorities seemed unable to find an internal solution. Attitudes worthy of colonial clichés emerged: experts declared it necessary to import knowledge "from other places and other times" (Metzger, 2009), as if solutions could not be found locally.[39] The authorities did not seem to have grasped the issue of legitimacy with reference to indigenousness. To quote Christian Morel (2002), the decision to conduct a massive mosquito control campaign seemed "absurd"; mosquito eradication is a "classic technocratic model" (Morel, 2002, 216), a "techno-science"

[39] "This form of blindness therefore raises the question of the conditions in which medical knowledge and clinical experiences from other places or other times should be mobilized" (Metzger, 2009, 192).

solution that makes no concessions to existing mosquito management practices. Such practices were hardly even mentioned. The urgency of the situation led to hasty solutions, cobbled together as if the island had never had to deal with mosquitoes before. DRASS press releases did not mention local practices (e.g. mosquito nets, smoke, plants such as lemon grass and geraniums), or rather they were hidden behind vague advice such as, "Do what you can to protect yourself effectively." The way the authorities phrased the epidemic resulted in a denial of any previously acquired knowledge and had the prefect declare, "Reunion islanders are going to have to live with mosquitoes" (the prefect quoted in, *Jir.* 15.02)!

At this point, several registers of uncertainty coexisted, and the breakdown of familiarity (Borraz, 2009) made choosing the appropriate solution impossible (the undecidability described by Chateauraynaud, 2008). Soon, however, this uncertainty became a serious anxiety (the third register of uncertainty) as new, more acutely vulnerable figures entered the picture. The TOS strike served as a reminder that the start of the school year was approaching and with it an influx of potential new victims. Without necessarily being infected with chikungunya, school children and staff were considered doubly vulnerable: to the epidemic on the one hand and to the solutions used to combat it on the other. The press began to describe mosquito eradication as the widespread "flytoxing" of Reunion, initiating construction on the media stage of the environmental wrong produced by mosquito control.

> For a few weeks now, school yards and garbage dumps have been bombarded, flytoxed, and sprayed hand over fist. There is also a lot of poisoning; there are children with irritations, a teacher collapsing here and a pesticide sprayer with red eyes and inflamed skin there. Not to forget, of course, the impact of this treatment on the flora and fauna. Bees – the first link in an ecological chain now affected across the board – have already paid a heavy price. Even though the departmental council and some municipalities now refuse to use Fenitrothion, Temephos and other pyrethrins, DRASS 4x4 nocturnal commandos continue to spray synthetic insecticides in urban areas and green spaces. (*Jir.* 15.02)

3.2 From fearsome flytoxing to safer control

"Flytoxing" became the generic term for mosquito control. Often used with a negative connotation, the term served to criticize and caricature the application of pesticides. A whole new vocabulary developed around the term: a flytoxed child, a flytoxed dog, a flytoxed "*endormi*" were expressions used to describe the voluntary weakening of the most vulnerable.[40] Pesticide sprayers became known as "flytoxers", making it impossible to know whether they were army, DRASS agents, or municipal, regional or departmental employees. The vulnerability of flytoxed victims reflected that of the "novice" flytoxers themselves, whose lack of training smacked of an organized poisoning of the population. Photographs of "nocturnal flytoxers" dressed in hazmat suits symbolized the danger. The fact that they operated at night reinforced the feelings of vulnerability surrounding their unwelcome visits, while the hazmat suits were a further reminder that the insecticides used were toxic.

Flytoxing was the word to crystallize all fears. The crisis unfolded against a backdrop of controversy. People sought an ideal culprit to blame for the disorder and chaos. Other grievances were voiced, particularly concerning the pesticides, and contributed to the construction of an environmental wrong caused by mosquito eradication. The controversy over the chemicals' toxicity remained in the headlines throughout the period studied. It was presented as the debate of experts and *foreigners*, highlighting the twofold distance of these actors: they were distant in status and distant in identity. The mosquito control campaign seemed the product of collusion between experts, who supported the use of the pesticides, and the state, which arranged for the products to be sprayed despite being fully aware of the eco-centred misgivings.

In the conflict between state institutions and local authorities, the former were at an immediate disadvantage. They were considered to be symbolically distant from local issues, because of what was seen as the excessive recruitment of state officials seconded to a territory that they supposedly approached as a former colony. "*4 chercheurs débarquent*" – "Four researchers arrive" – wrote the *Jir.* thus lumping experts and state officials together in the category of "*débarqués*" or "outsiders" (*Jir.* 15.02, p. 18).[41] This presupposition that some were close and others symbolically distant afforded a certain indigenous capital to local actors, which automatically upped their approval rating and served to mitigate their responsibility in the crisis. These contextual effects discredited in

40 *Endormi* or "sleeper" is the vernacular name given to the chameleon panther or *furficer pardalis,* "A species originally endemic to Madagascar. It was introduced many years ago to Reunion Island where, although exotic, it benefits from the status of a protected species" (source: parcsnationaux.fr).

41 The pejorative term "*débarqué*" was regularly used to indicate the non-indigenous origin of newcomers who had just arrived on the island.

advance anything outside experts had to say. Their words were stigmatized before they could be heard. Because of their origin, these experts lacked the environmental capital of indigenousness (Bouet *et al.*, 2016) they would have needed to compensate for the doubts amply fuelled by the controversy. In other words, the experts were considered to have failed to make an effort to "convert" their expertise to the register local populations expected (Bouet *et al.*, 2016).

According to the press, the government had organized the widespread use of two organophosphorus pesticides, Fenitrothion and Temephos, in 2005. The prefect ordered these products in late 2005 and the orders required for their application were published in the *Journal Officiel* on 15 December 2005. Fenitrothion and Temephos are synthetic chemicals. One kills adult mosquitoes and the other kills mosquito larvae. Together they are supposed to be the most effective on the market, yet they got very bad press. Non-Reunionese experts considered the widespread application of these products a necessary evil to "destroy the adult mosquito population" (*Quot.* 08.02). Faced with the enormity of the health issue, the experts had approved the choice of product based on an "efficacy/toxicity ratio" (*Quot.* 08.02). However, when they tried to provide precise responses to the complaints over the use of these two pesticides, the experts were unconvincing, even with the authoritative argument of potential child deaths. They swept aside suspicions of the pesticides' harmfulness, saying it was enough to use the right amount. In the early days, no mention was made in the press of the precautions for use; the "correct dosage" alone was recommended (*Quot.* du 08.02).

Some experts acknowledged a cruel lack of data on initial mosquito populations. They were unable to measure either the impact or the effectiveness of the mosquito control operations. "We are not focused on measuring impacts because this is an emergency situation" (*Jir.* 02.03, p. 16). Here, the health emergency argument was used to make up for the "uncertainty" (Chatcauraynaud, 2008) of environmental decisions. At the same time, environmental experts challenged the effectiveness of mosquito eradication and predicted another catastrophe which, if not more urgent, would be more lasting. Two competing impact measures came into play: environmental impact versus the impact on reducing the epidemic. Justifying or criticizing mosquito control seemed like a matter of faith, even for the experts, prompting *Jir.* to say that there was nothing for it but to devote oneself to "Saint Expeditus and Saint Flytox"

(*Jir.* 02.03, p. 16).[42] Moreover, the scientific doubt about the health impact of mosquito control did nothing to benefit the experts, who seem to have underestimated the popularity and local reach of ecological issues. Their indecision only fed the concerns of the population, who used the principle of precaution to argue against the pesticides chosen by the authorities.

The controversy arose from the discrepancy between the substance (a growing interest in "nature" that experts did not seem to see coming) and the form (environmental management considered catastrophic for both the environment and human health). An "ecological disaster" seemed imminent; no matter how much the prefect denied it – "I am not poisoning Reunion" – the damage had been done. The "chemical tsunami" turned the management of a health emergency into a controversy pitting the autocratic power of local authorities against popular disapproval of their management methods. Then, with the long-awaited arrival of BTI, the controversy seemed to fade. The prefect announced its forthcoming delivery at the same time as the second control plan. The reasons for the delay in ordering BTI (including the cost of the product, which is more expensive than other larvicides) were not indicated. However, according to the experts interviewed by the press, the effective duration of BTI was insufficient and it would require repeated applications, unlike Temephos, which would provide "two months respite" (*Quot.* 08.02, p. 15). But BTI had a good reputation among the population; it was a "bio-pesticide" symbolizing safer mosquito control. As early as December 2015, the Duhamel report referred to the media's presentation of BTI as "biological" and questioned the way this favourable portrayal of BTI was being constructed (p. 42). By querying the criteria underlying this categorization of BTI, the report's authors showed in advance that other products would be controversial. Public support for BTI, with its bio-pesticide label, demonstrated the desire for "safe" eradication as opposed to the "polluting" eradication organized by the state.

In mid-March, *Le Quotidien* (14.03) reported that the control plan had learned from its mistakes and was "much safer". A mosquito control team was working during daylight hours, local door-to-door mediators outnumbered the spraying teams (according to the article, there were four mediators for every sprayer in a hazmat suit), a bio-pesticide (BTI) was

42 Saint Expédit is the saint of lost causes, to whom the Reunionese devote a remarkable cult with small red altars that line the island's roads.

being used with restraint and animals taken into account, and domestic practices were no longer stigmatized. The controversy abated. Both the grievances and the strong popular desire for safe mosquito control seemed to have been heard. Did this desire have an ethical source or was it just a consequence of ad hoc arguments used in an environmental controversy? Environmental spokespeople – some green councillors but mostly from the non-profit sector – had sent a clear message from the beginning. They had been condemning the widespread use of harmful products for a year already. In contrast, the stance of other elected representatives or citizens' groups was more ambiguous. In the calls for safe mosquito control, a hotchpotch of different desires for naturalness were expressed.

From the population's point of view, the "chemical tsunami" (front page of *Jir.* 13.02) fuelled widespread distrust of the "chemical technosphere". Such appeals corresponded to a general trend for "natural" treatment practices and "green" remedies (*Quot.* 07.02). Public support for "natural" products grew stronger as industrial products and repellents were increasingly perceived as ineffective. Naturalness appeared to be an alternative solution to all-out chemicals and would enable a focus on the environmental footprint of mosquito control. It was a matter of looking beyond the crisis to a post-epidemic period in which the impacts on biodiversity and groundwater would have to be dealt with.

For management, the pollution caused by widespread mosquito control was simply collateral damage (*Quot.* 9.2.6). Opposing them, numerous environmental groups made their voice heard, including ecological associations such as the Ecological Generation Movement of La Reunion, *MGER* (*Jir.* 12.02.06), entomologists, local elected representatives and citizens' groups created in response to the chikungunya epidemic. Their repeated warnings depicted the environment as the victim of a "blind flytoxing" policy (*Jir.* 2.3, p. 16). Some media reports also anticipated the long-term consequences of such massive use of chemicals. They gave these groups the opportunity to raise the issues of groundwater pollution, the consequences for beekeepers and difficulties for organic farmers.

Some press articles let the ideology of "treating nature with nature" be expressed in support of biological control, such as the introduction of new predatory species to combat mosquitoes – "The kind of proposal that garners the wrath of scientists" (*Quot.* 04.02) – or using entirely natural treatments.

Yet the calls for biological control cannot be explained by concern over the pesticide treatments alone. This concern did not extend to all the products used to kill or repel mosquitoes. The arguments for safe mosquito eradication did not question individual mosquito control practices. The root of this concern should be qualified, as it was clearly expressed only with regard to *massive* mosquito control operations. Does this make it NIMBYism? Can we see in the demand for "safe control" an attempt to overcome a sense of powerlessness over a mosquito eradication programme that was beyond citizen's control?

While the toxicity of the products being sprayed en masse was worrisome, little attention was paid to the ingredients in domestic mosquito control products, especially given the calls to make personal protection more accessible. The idea that repellents (sprays, electric diffusers) should be distributed for free was put forward in the name of social justice. Councillors called on the state to finance repellents for the most disadvantaged. The press criticized a "deliberate" shortage, which the authorities and pharmacies denied; increased demand for repellents was supposedly making profits for stockists, who were therefore ensuring the cheaper products were out of stock to force customers to buy more expensive brands. "Supermarkets bring out the armada" (*Jir.* 19.01). Pharmacists and supermarkets were widely accused and referred to as "war profiteers" (interview with Virapoullé, mayor of Saint André, *Jir.*). It was not until the beginning of March that domestic repellents came under scrutiny, but this essentially concerned their ineffectiveness and caused no real debate over their toxicity.[43] Again, a lack of transparency was highlighted.

> It is difficult to evaluate the impact of an insect repellent or its dangers because no long-term studies have been carried out [...] in France, repellents fall under cosmetics and personal hygiene legislation which is not very restrictive. No proof of effectiveness is officially required, and the only compulsory tolerance tests are carried out on animals. (*Jir.* 14.03)

The public was advised to seek pharmacists' recommendations or to rely on their own resourcefulness, even though the press releases being put out called for increased personal protection. How could people choose the "right product" when the authorities' press releases only specified

43 Which led an expert to claim that insecticides for everyday use at home caused more pollution than the mosquito control campaign (*Quot.*, 08.02).

"repellents"? The Departmental Council advised people to "protect themselves against mosquito bites using electric fans and diffusers, to use repellent creams on babies and young children and for adults to regularly use a spray" (*Quot.* 30.01). The recommendation was to adapt individual protection to the "target" (child or adult). DRASS was vaguer still: for personal protection, people should use a repellent (*Quot.* 13.01), wear long, pale-coloured clothing when mosquitoes were present and be especially careful at sunrise and sunset.

The controversy brought environmental ethics to the fore and turned the environment into an object of heritage in itself. Its intrinsic value (the protection of biodiversity) was closely linked to respect for environmental health and justice. Such ethical issues stemmed from the introduction of a long-term perspective over the course of the epidemic. "The war against mosquitoes is a long-term effort, not a blitzkrieg" (*Tem.* 07.03, CDL). A sustainable resolution to the chikungunya epidemic would take time.

Calls for "biological mosquito control" stretched beyond the urgency of the epidemic to a long-term perspective. References to the term "biological" became almost like an incantation. "Bio", which in French also means organic, often served as an ideological catch-all: "There is a relative lack of clarity about these subjects both in terms of concepts and of what it really means to adopt these forms" (Lamine *et al.*, 2010). But public support for BTI did not wait for the controversy over "flytoxing" and "children falling like flies in mosquito-treated schools" (*Jir.*). It was voiced from the outset in aid of a broader cause: "Enough poison! Biological control, and fast!" (*Tem.* 03.01, p. 3). In an article dated 3 January, *Témoignages* criticized the use of organophosphates, "one of the most toxic pesticides for humans and animals." So, calls for biological control began early, long before the epidemic declined (*Tem.* 03.01; *Tem.* 26.01). The health crisis only made them more visible. Environmental issues did not compete with human issues; quite the opposite. Moreover, the social context was considered wholly unconducive to the prudent use of these pesticides. It was therefore in the name of social and environmental justice, *Temoignages* argued, that safer measures were required.

> We are constantly reminded that out of 780,000 inhabitants, 110,000 adults are illiterate, but in order to make this type of product available, the authorities act as if everyone can read, write and above all understand the pictograms and warnings on the labels of these poisons. (*Tem.* 03.01)

The alarm raised over mosquito control formed a rhetoric that began with a question, "Why rule out biological control"? (*Tem.* 26.01), and ended with a demand: biological control must be used in the name of social justice. Protection of the environment was presented as the key to protecting human health. So what solutions were proposed? The press expounded a number of initiatives over the course of the controversy, but we will focus on just one: cleaning up the island. Clean-up activities involved solidarity on a human scale. It was thought that a series of concrete, manageable individual efforts added up together could really make the difference. Furthermore, these measures based on environmental ethics represented an "endogenous" community response. They illustrate the second part of our argument: the health crisis was a movement for environmental justice.

3.3 From environmental risk to wrong: the chikungunya crisis through the lens of environmental justice

These environmental ethics were not formed in response to the management methods of health institutions. They originated in citizen mobilization concerned with environmental justice. But they should be resituated in the political game of pass the buck that saw institutions blaming each other for the epidemic. For the institutions responsible for public health and safety (such as DRASS and the Prefect), mosquitoes were Enemy No. 1 (*Jir.* 05.01) and the environment remained a major threat. DRASS aimed to inform the population of this threat via "environmental mediators" working on the ground to raise awareness about the dangers of mosquito proliferation (Dupé, 2015). This mediation was "environmental" in name only; it was not driven by "ecological" concern. The mediators' "environmental" mission was to inform people about the environmental dangers that threatened inhabitants in their immediate surroundings (e.g., saucers of water under potted plants and rainwater recovery vessels as potential breeding sites, etc.). It was not their responsibility to foster an ecological consciousness that would translate into respect for the environment (Bourg and Fragnière, 2014). In contrast, local and regional authorities already had an awareness of the environment as a common good and heritage resource. They expressed concern for its protection and were instrumental in making the environmental health issue explicit.

The institutions' different representations of the environment appear to be structural and to derive from the missions of each institution. It seems logical that DRASS, a public health institution, developed a human-centred ethic while local institutions with environmental "powers" developed a more eco-centric ethic.[44] As Elise Mieulet and Cecilia Claeys have argued (2016), such divergent representations run the risk of "polycephalic" and potentially contradictory environmental policies. The press seized on these contradictions and used them to turn the institutions against each other. Indeed, the narrative of the clean-up activities must be resituated in the context of a power struggle between the state and local authorities, where both sides blamed each other for the health crisis and their supporters demanded that guilty heads roll. On 4 January the prefect "self-assuredly" announced the chikungunya control plan (at the same time as the mosquito eradication operation). He gave an update (*Jir.* 5.01) and assessed the state of the "preventive health battle" by announcing the results of the Saint Paul clean-up programme: "Householders have removed 550 tonnes of waste from their yards."

> I will write to the mayors about this because my offices have listed more than 200 illegal dumping sites which are all areas of infestation. And I will make the army available to help with the most difficult sites. (*Jir.* 5.01)

At stake in these operations was the coherence of the chikungunya control policy: "It is a matter of developing as coherent and comprehensive a control strategy as possible" (*Tem.* 14-15.01). Right from the start of the crisis, yard cleanups were one of DRASS' recommendations.[45] But big clean-ups were not an initiative specific to mosquito control. After a cyclone, it is customary for communities to clear the island of waste and debris deposited by torrential rains. The island's major clean-up

44 According to article 72 of the French Constitution, "Territorial communities may take decisions in all matters arising under powers that can best be exercised at their level". Thus, municipalities, *départements* and regions organize and operate many public services that the law has entrusted to them (source: http://www.collectivites-locales.gouv.fr/competences). The Environment Code and the Local Authorities Code provide a framework for the eleven sovereign and/or mandatory powers of local authorities, including "social action and health" and "the environment".

45 A DRASS statement recommended three courses of action against chikungunya: 1. Protect yourself. 2. Clean up. 3. Keep informed and report any problems (*Quot.* January 2006). The Departmental Council reversed the order in its press releases: 1. Eliminate mosquitoes from your garden 2. Protect yourself against mosquito bites (*Quot.* January 2006).

operations started up again in 2006, with the prefect pointing the finger at the local authorities, whose responsibility for environmental affairs should also include waste management. They were openly accused of laxity. Certain associations took up the rebuke. SREPEN called attention to the structural backlogs accumulated over several years; the lack of a composting and recycling station had led to landfill sites quickly becoming saturated. "Things are getting worse, local authorities are prevaricating and ducking responsibility" (*Tem.* 07.03). The mayor of Saint Louis, a town sadly dubbed "the chikungunya capital", announced an ambitious mosquito control plan and his intentions for a large-scale clean-up:

> We will settle our scores with the government later and even if these measures fall outside our remit, we must pool our know-how during this crisis situation. (*Jir.* 17.01)

The clean-up operations were an opportunity for governing bodies to air each other's dirty laundry in public. An "inverted" competition ensued, aiming to stigmatize the most negligent municipality. The state and local authorities feared that their respective "negligence" would earn them the role of main culprit in the spread of the epidemic. The extensive publicity for "yard clean-ups" made it appear as if municipalities were taking part in the "mosquito war" effort on their own initiative. They pretended to ignore that waste management had always been their responsibility. Justifiably, since these authorities changed nothing about their "usual" waste management practices. They even complained that they had been given no directive to do so. Yet a problem soon emerged. The calls to "clean up your yard" led to increased use of landfill sites (*Quot.* 11.02 "Landfill sites at full capacity"). Newspapers reported on landfill overload (*Jir.* 30.01 "Hermitage landfill site overflows") and even saturation (*Quot.* 09.02 "Le Port composting station nears saturation point"). Now, in addition to hospitals overwhelmed by the arrival of sick patients, the island had to deal with overflowing landfill sites. The yard clean-up campaign was a victim of its own success.

Saying that landfills were full was risky: it could either be an acknowledgement of past negligence – negligence that environmental associations had been criticizing for years – or place the blame on users, stigmatizing them for their "filth". Reunion Island "produced more waste than sugar cane" (Mayor of Saint Louis, *Jir.* 17.01). In this game of pass the buck, however, the idea was to avoid raising too many questions about individual responsibility which would only add to the crisis.

Yet the reminders about fines for failing to observe collection dates did stigmatize domestic waste management practices (*Quot.* 25.03). Some institutional communication also made stigmatizing remarks concerning VC.[46] In the end, the "war on waste" (*Jir.* 17.01) did not hesitate to point the finger at the islanders' lack of hygiene.

When the time came to assess the waste management "crackdown", the press reported a request from the prefect to double the frequency of garbage collection and deal with some 500 illegal dumping sites. But in the game of pass the buck, some municipalities came out looking good. To deal with the mosquito control plan, they highlighted their past experience with VC, which was not limited to reducing the spread of chikungunya. Not all landfills were "overflowing", as the technical director of CIVIS (Intermunicipal Community made up of the municipalities of Saint-Pierre, Saint-Louis, L'Étang-Salé, Petite-Ile, Cilaos and Les Avirons) boasted, for example:

> CIVIS had an undeniable advantage over the island's other councils: it had always been a few steps ahead in its waste management. (*Quot.* 11.02, quoting the technical director of CIVIS)

It should be noted that environmental associations had long propounded clean-up operations as a health necessity. The objectives of this "clean and tidy" strategy were detailed in the press releases of a long-standing Reunionese environmental association, SREPEN. Local authorities took up this eco-centric argument in unison during the crisis, but it seems to have been used in the first instance by the municipality of Le Port. Le Port Council was depicted in the press as a trail-blazer based on its prior initiatives to combat dengue (*Jir.* 25.01). It was also the only city on the island to have stopped using Fenitrothion (2004) and replaced it with BTI and K'Othrine (deltamethrin) (*Quot.* 11.02). Refusing to insult the population with the assumption that it "did not know how to sort waste" (and thereby keeping its voters happy), the council published a brochure entitled "Stop Mosquitoes" which was distributed to all households along with a comic strip for children.[47] Use of BTI,

46 A DRASS press release stated, "I respect nature. I only throw waste in the places provided specifically for this purpose (waste collection centres, etc.)", thereby attributing the unsociable practices of a few people to the entire population.

47 "The council is thinking about the future, so aims to raise awareness among children especially. That is why it put out 15,000 copies of a comic strip called 'Tik Tak Chik'" (*Tem.* 08.04).

community action, clean-up operations and targeted communication – especially aimed at children who were considered crucial in spreading the word – all these actions were motivated by an "ecological concern" that was identified from the outset as a key issue.

> Yesterday, Le Port Council reviewed the spread of the chikungunya epidemic in the maritime city, the results of the mobile waste disposal experiment and the necessary coordination between the council teams and the brigades sent by DRASS. In all this, the use of suitable products is crucial to avoid turning the current economic and health disaster into an ecological disaster as well. (*Tem.* 08.02)

One argument seems to have prompted Le Port council to innovate when dengue was first reported in 2004: "Ecology against chikungunya" (*Jir.* 25.01) was associated with a social justice policy. Mobile waste disposal operations were added to the range of community-based mosquito control activities already in place before the chikungunya crisis. The initiative involved mechanically removing mosquito breeding sites (namely organic, metal and bulky waste) using a lightweight mobile collection system (*Tem.* 08.04).[48] The objective was threefold: 1) to remove waste from private yards and allow effective spraying to take place; 2) to prevent "trucks heading to landfills without their waste being sorted, as they usually do" (*Quot.* 09.02) and 3) to have the waste collectors report new cases of chikungunya, thereby improving health monitoring and enabling areas to be treated again when necessary. No matter how important it was not to stigmatize people for the way managed their yards, the public did have to be taken to task about illegal dumping. To counter this "real societal problem in Reunion", plans to manage illegal dumpsites were presented as solutions for greater environmental justice; it was a matter of ensuring preventive measures benefited the poorest. These inhabitants lived closer to illegal dumpsites and were potentially more exposed to viral reservoirs and infected persons (Taglioni, 2009).[49]

---

48 "Since 26 January, mechanical mosquito control has intensified with an experiment in 'mobile waste collection'. A mobile 'yard clean-out' operation conducted in every neighbourhood has enabled collection of a quantity of green or bulky waste equivalent to nearly two thirds of the quantity collected at the same time (from 26 January to 21 February) by traditional (unsorted) collection. […] The lightweight mobile collection system for green, metal or bulky waste will be maintained" (*Tem.*, 08.04).

49 François Taglioni (2009) has shown that urbanization is a risk factor for the transmission of arboviruses in Reunion (Taglioni, 2009a).

3.4 Decolonizing health and environmental crises

Figure 3. *Illegal tire dump, band'cochon.*

Photo credit: www.bandcochon.re

Tires were a powerful symbol that was frequently used in the press to illustrate the need to clean up the island. The tire is an artefact with a strong symbolic charge. Tires bring to mind the abandoned wreck that has had its wheels removed and repurposed in the garden. The use of tires as outdoor flower pots – known locally as macottes – was problematic since they have no recipient underneath which can be emptied and make an excellent mosquito breeding site.[50] Tires piled up in illegal dumping sites allude to a colonial cliché stigmatizing the islanders' so-called "lack of hygiene". Tires are also the key to transport. Consequently, they symbolize the "cult" of mobility, attributed with every virtue in Reunion: mobility is seen as important for social success and critical for economic success (Simon, 2010). Mobility between an island and mainland France – which islanders continue to call the "Metropole", despite having become a *département* – is crucial, but this mobility also carries diseases

[50] *Macotte* is the Creole word for a makeshift flower pot.

from the South to the North (*Jir.* 12.02) and spreads diseases between countries of the South. Tires represent the overly pervasive automobile industry, producer of unmanaged waste made visible by illegal dumping. Tires also inadvertently transport mosquitoes, allowing them to cross borders undetected, thus demonstrating their invisibility and consequent power to cause harm.

Talk of tires raised the spectre of the economy: an economy that bore the marks of exclusively automobile-oriented thinking; an economy so fragile that it was impossible to talk about sensitive issues such as the need to preserve territorial continuity at all costs. What issues were not raised by the mention of tires and car wrecks, the only visible signs of an automobile-centred development policy?[51] The press, usually so quick to flush out the hidden aims of its interlocutors, wrote of car wrecks as either proof of unclean, uncivil islanders, or of a system which has no means of getting rid of them. But the economic system at the origin of it all was rarely called into question. Popular practices were stigmatized without reference to their systemic causes: the car industry lobby, economic monopolies and road infrastructure that offers little in the way of public transportation.

These "things left unsaid" can be interpreted as so many unthinkable thoughts about the health crisis in a non-emancipated postcolonial context. It seems to have been impossible to say that humans constitute a reservoir of infection and to point to people's modern habits as guilty of spreading the disease. On 12 February, *Le Quotidien* reported on three cases of chikungunya in Martinique following a tourists' visit to Reunion and, rather than directly calling human mobility into question, blamed the spread of mosquitoes. "Mosquitoes spread through the eggs they lay […] in tires" (*Quot.* 12.02.06). Evoking the risk caused by the free movement of people would make them a virus reservoir (Gaüzère, 2006) and potential vectors. Such causal attribution would justify preventive measures that seemed to hark back to a pre-modern era. Any limitations to human mobility attracted fierce condemnation. When Reunion Islanders were denied a hotel booking in Paris as a "precaution" (front page of *Jir.* and *Quot.*), this exclusion was castigated as racist.

51 Analogy to the Bourdieusian formula, "In order to develop the theory of rites of passage any further, one has to ask the questions that this theory does not raise" (Bourdieu, 1991, 117).

Whereas the movement of mosquitoes was a subject of obsession, limits on human mobility were never decided upon or even suggested. This option remained in a causal blind spot, as if it was too frightening to consider. Pointing to the effects of the free movement of humans as a factor in the spread of the epidemic would have challenged the economic model of these islands centred on consumerism, tourism and imports (all dependant on freight transport) (Taglioni *et al.*, 2009). "Mosquitoes transmit disease and disease is dangerous, *ergo* mosquitoes are dangerous." The simplicity of this sophism prevented other ways of thinking from emerging (Watin, 2009). The chik' crisis was environmental in more ways than one: as a mosquito problem rather than a human problem (the environment is a threat) or in terms of the environment as a heritage to be preserved. In both cases, the environmental dimension served a purpose: to avoid questioning the human practices that made the crisis possible.

The authorities were silent on all health solutions that did not regard the environment as a threat or that considered human movement to be a risk factor in the epidemic. Yet this factor had already been mentioned by Professor Gaüzère in his 2006 book on the chikungunya epidemic. Blaming the movement of mosquitoes prevented human movement from being seen as a vector for mosquito movement. During the three months covered by this study, human mobility was not targeted in the management of the crisis. On the contrary, the spectre of a fragile economy threatened by dwindling tourism led to lamentations over cancelled bookings, the reduction in flights and even the decline in the number of people visiting public gardens.

[photo 2] *Souch's drawing on the cover of Idelson and Legeden, 2011*

Any "break" in territorial continuity was seen as an additional tragedy. Condemning such a break served in passing to accuse potential culprits. In the *Jir.* on 13.02, the mayor of Saint Paul criticized the prefect's incompetence "which allowed the virus to enter Europe." Yet at no time was the suspension of human mobility put forward as a potential solution for halting the spread of the virus. Conversely, some even hoped the disease had spread to France (CDL), as if the spread of the disease might also increase empathy for the sick Reunion Islanders left to fend for themselves.

Quarantine practices were occasionally hinted at. The state required new cases of chikungunya to be reported. This was so the infected households could be treated for mosquitoes to avoid any further contamination. And yet it was also recommended *not* to isolate patients, so that they could be monitored, to avoid the potentially deadly effects of isolation and, especially, to free up room in the overcrowded hospitals. As the quarantine of patients was collectively and ethically impossible, personal protective measures were essential. It is easy to understand why the authorities insisted that "each individual is responsible for his or her situation" (quote from the prefect, *Tem.* 05.01). Yet this order was barely tenable in a context where the persistence of pronounced social inequalities prevented reasoned pesticide use and limited access to modern care. Individual protection measures seemed a euphemism for quarantine. Even when the disease spread internationally, no reference was made to limiting mobility. Only the recommendations (to wear long clothes, etc.) remained: "to protect the people around you, (you must) first protect yourself" (*Quot.* 12.02.06).

Faced with a lack of national sympathy regarding the health crisis, four Reunionese senators addressed the Senate. "We are not here to beg" (Vergès, *Quot.*) for national solidarity for a disease that "has spiralled out of control because it affects a developed country in a tropical region" and for "an ecological catastrophe due to mosquito control products" (Gélita Hoarau, *Quot.* 08.03). Contagion could not be spoken of, as it would risk breaking an already fragile physical and symbolic territorial continuity. The press even built up popular empowerment strategies as being motivated by the feeling of an unjust breakdown in national solidarity. According to the *Jir.* of 27.01, the association "*Les Réunionnais contre la chikungunya*" was formed following a telephone altercation between its future president and a member of Ministry of Health staff in Paris. It became important for her to collect testimonials from sick islanders and centralize missing information, arguing "we are not sub-French". The editorial the following day highlighted the accusations of incompetence (or "structural vulnerabilities", Watin, 2009), claiming it was not so much the state as its "thoughtless and transient" local representatives who had organized the silence. "[The prefect] will change but [the Reunionese] will remain" (*Tem.* 07.03).

References to indigenousness came up against accusations of local leaders' complicity and laxity. Indigenousness was racialized ("the white leaders from Paris"; "mosquitoes will bite everyone: yabs [local working class whites]… even reds"). References to origins granted a certain indigenous capital to those involved in the crisis, raising public approval for some actors and mitigating their responsibility, while working to delegitimize others. Some believed that "colonized minds" expressed "Northern common sense". In any case, it was the "North" that managed the epidemic in this "Monaco of the Indian Ocean". "Southern common sense" was disdained as popular imagination not worth describing, let alone building on. On the contrary, the population was stigmatized in advance, made responsible for the spread of the epidemic and referred to in terms of clean and dirty (such as in DRASS press releases). "Why are common sense, practical, inexpensive solutions not discussed" (*Jir.* 04/02) to isolate patients and prevent chikungunya from becoming a "nosocomial neo-risk" liable to spread the infection in hospitals?

Conclusion

The chikungunya epidemic was an environmental crisis from the start. Already in early 2006, the press placed environmental concerns on the same footing as other factors as a key element in resolving the crisis: "Chik, a social, sanitary and environmental problem" (*Tem.* 02.01).

This concern for the environment was due to two factors. The first was a huge protest movement against the prefect's new mosquito control plan. The protest movement followed the twists and turns of the controversy over the pesticides' toxicity. Children and animals were presented as the victims of mosquito control. Their vulnerability served to strengthen the construction of risk surrounding the "solution" to the epidemic. The environment here was not seen as "heritage" – it was "simply" useful and had to remain so, without fear of being spoiled by the misplaced efforts of "flytoxers". While the health authorities made the environment's heritage- and use-value into two competing issues, this first movement removed the barrier between environmental management and health management.

The burgeoning heritage value of the environment was made visible by the controversy. As community actions proliferated and the perspective became more long-term, "nature" was constructed as an asset to be passed on. In criticizing the aggressive and disproportionate mosquito control strategy, the local authorities developed an appropriate pro-environment argument. They may have overplayed the environmental ethics card in the game of pass the buck, but by justifying the actions they were taking "against" those advocated by the state, they allowed other senses of justice to be heard.

The Duhamel report had predicted that the products used could become a controversial issue (Duhamel *et al.*, 2006). The press took this idea and ran with it. With no response forthcoming from the administration, the silence seemed designed to maintain ignorance. The construction of the "mosquito control risk" arose out of an accumulation of intolerable factors (Bourdelais and Fassin, 2005). The imposition of a control plan that was "stronger than the disease" (*Jir.*) became intolerable. It became intolerable that people were prevented from protecting themselves due to the shortage of repellents. It became intolerable that the collective prevention of an epidemic in a crisis period should hinder the protection of the environment and endanger its future. In addition to mistreating patients and failing to prevent the spread of the epidemic, management of the health crisis now threatened the future. It was under these conditions that concern over the environment was made public.

Yet advocating for safer mosquito control and rejecting the extensive use of chemical pesticides did not necessarily mean opting for a just, "moral" solution that would guarantee a balance between human and environmental interests. Like chemical control, the "biological" control favoured by the press required the mosquitoes to be killed; the eradication programme "must take no prisoners" (*Jir.* 15.02). As early as January 2006, however, an ecological catastrophe was predicted in the press. This ecological awareness raised the question of the role of mosquitoes and insects affected by insecticides in the food chain. Was it possible to limit the spread of the virus other than by eradicating mosquitoes? What were the long-term consequences of massive mosquito control? "Killing colonies of insects" (*Quot.* 03.01.06) was gradually called into question, especially as another factor in epidemic mortality became increasingly clear – the poor health of the population (front page of *Quot.* 08.02.06). Mosquito eradication came to appear only a partial and biased means of controlling the epidemic. This second ethical movement brought "nature" to the fore as a common good battling with a moral issue and no longer simply a managerial one. It led to a reframing of the health problem: environmental protection became the key to human health.

There were two conflicts running through the environmental issue: the tension between the North and the South and the tension between the environment and health.

The chikungunya epidemic triggered the break between a state (outside) capable of blindly imposing an environmentally destructive policy, and local authorities (inside) willing to blow the whistle on an impending "ecological catastrophe". The press played out the dispute in its pages; the environmental cause was like a child caught between two sides in a messy divorce, tearing each other apart through the press. The fact that the health authorities framed the crisis in environmental terms (Metzger, 2009) saw the population clamouring via the media for humans to be put back at the heart of the agenda.

The people of Reunion Island, through vehement complaints and demands relayed by the press in shock headlines, seemed to express their sense of abandonment in a single voice. The population countered the involuntary, disorganized environmentalism of the health authorities with a frenzied, ecological humanism. To be convincing, this humanism played on its legitimacy to stigmatize the health authorities, which were segregated into a symbolic "inhumanity". References to indigenousness brought those representing the "outside" into opposition with the patients "inside" and rendered unjust the authorities' orders to take individual responsibility. This tension has already been addressed in the literature on the health crisis (Taglioni, Watin, Idelson and Metzger, 2009).

Our findings show another dividing line: between massive mosquito control, constructed as an additional risk during the epidemic, and individualized protection. The mosquito control risk underscored the vulnerability of children and animals. The chameleon, an emblem of animal vulnerability as well as of Reunion, attested to this growing interest in the environment. It may seem paradoxical (or immoral) to look toward nature and the long-term when humans were suffering. However, such attention to nature crystallized existing tensions: the mosquito was the tree that hides the forest. Public health authorities were so focused on mosquitoes that they appeared to have forgotten humans and the multitude of interests and values that might lead them to protect the environment. Nonetheless, the crisis had to remain environmental to hide its economic motives: this crisis was also that of a former colony. Colonial clichés of clean and dirty having caught up with them, the authorities failed to overcome their lack of legitimacy and adapt their practices to the postcolonial island context.

References

Dupé S. (2015), *Séparer les moustiques des humains à La Réunion. Co-production d'un nouvel ordre socio-naturel en contexte post-colonial.* Thèse en anthropologie sociale et ethnologie, Université de la Réunion.

Dupé S. (2015). « Transformer pour contrôler », *Revue d'anthropologie des connaissances*, 9(2), 213-236.

Duret P., Cubizolles S. and Thiannbo M. (2013). « La crise sanitaire du chikungunya : une épreuve de recomposition des rapports sociaux à La Réunion », *Sociologie*, 4(3), 317-332.

Fassin D. and Bourdelais P. (dir.) (2005). *Les constructions de l'intolérable : études d'anthropologie et d'histoire sur les frontières de l'espace moral.* Paris, La Découverte.

Gaüzère B.-A. and Aubry P. (2006). *Le chik, le choc, le chèque : l'épidémie de Chikungunya à la Réunion 2005-2006 en questions.* Sainte-Marie (Réunion), Azalées Éditions.

Gaüzère B.-A., Bohrer M., Drouet D., Gasque P., Jaffar-Bandjee M.-C., Filleul L. and Vandroux D. (2011). « Infection à virus Chikungunya à La Réunion en 2005-2006 : formes graves émergentes de l'adulte en réanimation. Chikungunya virus infection in Reunion Island in 2005–2006: severe emerging adult forms in the intensive care unit », *Réanimation*, 20(3), 211.

Gaüzère B.-A. (2011). « Hurler avec les loups ou crier au loup ? Ou l'art redoutable de la communication par temps de crise sanitaire », In B. Idelson et G. Ledegen (dir.). *Chikungunya : la médiatisation d'une crise* (169-186). Fernelmont, Éditions Modulaires Européennes et Intercommunications.

Idelson B. and Ledegen G. (dir.) (2011). *Chikungunya : la médiatisation d'une crise. Presse, humour, communication publique.* Fernelmont, Éditions Modulaires Européennes et Intercommunications.

Lamine C., Tetart G. and Chateauraynaud F. (2010). « Le bio comme reconfigurateur des controverses sur les pesticides et les OGM (1995-2008) », In *Actes La Réduction des Pesticides : enjeux, modalités et conséquences* (12 p.). Presented at Colloque Société Française d'Économie Rurale (SFER). La réduction des pesticides : enjeux, modalités et conséquences, Lyon, FRA (2010-03-11 - 2010-03-12). http://prodinra.inra.fr/record/39483

Metzger P. (2009). « L'épidémie de chikungunya : un problème de moustiques ? », In C. Gilbert et E. Henry (dir.). *Comment se construisent les problèmes de santé publique* (175-193). Paris, La Découverte.

Mieulet E. and Claeys C. (2016). « (In)acceptabilités environnementales et/ou sanitaires : dilemmes autour de la démoustication du littoral méditerranéen français », *VertigO-la revue électronique en sciences de l'environnement*, 16(1).

Morizot B. (2016). *Les Diplomates : cohabiter avec les loups sur une autre carte du vivant.* Marseille, Éditions Wildproject.

Rapport dit « Duhamel de la mission d'appui à la lutte contre l'épidémie de chikungunya à La Réunion », n° 2006/012.

Renault P., Sissoko D., Ledrans M., Pierre V. and Brucker G. (2008). « L'épidémie de chikungunya à La Réunion et à Mayotte, France, 2005-2006 : le contexte et les questions de surveillance et d'évaluation posées », *Bull Epidemiol Hebd*, 38-39-40, 343-346.

Setbon M. and Raude J. (2008). « Le chikungunya à la Réunion : facteurs sociaux, environnementaux et comportementaux en situation épidémique », *Population*, 63(3), 555-583.

Taglioni F. and Dehecq J.-S. (2009). « L'environnement socio-spatial comme facteur d'émergence des maladies infectieuses », *EchoGéo*, 9. https://doi.org/10.4000/echogeo.11168

Watin M. (2007). « Polémique, rumeur et tension : aspects de la crise du chikungunya dans l'espace public médiatique réunionnais », In S. Fuma et J. Chan Low (dir.). *Épidémies et pharmacopée traditionnelle dans l'histoire des îles de l'océan Indien. Actes du colloque international des 5-7 décembre à l'Université de Maurice* (241-251). Saint-Denis, Université de la Réunion.

Watin M., Metzger P., Taglioni F. and Idelson B. (2009). *Situation de crise, opinion publique et vulnérabilités. L'épidémie de chikungunya à La Réunion et à Mayotte.* Rapport final de l'IRD pour le ministère de l'Outre-Mer.

Yamada O. (2013). L'épidémie de chikungunya sur l'île de la Réunion, de 2005 à 2006. *Les Cahiers de la Recherche. Santé, Environnement, Travail*, 3, 12-14.

Gardens, pesticides and mosquito-borne diseases: an interdisciplinary comparison between mainland France and the French Antilles

Cécilia CLAEYS*, Valérie BERTAUDIÈRE-MONTES*, Christine ROBLES*, Magali DESCHAMPS-COTTIN*, Julie CARDI**

*Associate Professors, Aix-Marseille University, LPED,
**PhD Student, Aix-Marseille University, LPED

The goal of the PROLITENSAN[52] research programme was to develop an interdisciplinary protocol that combined ecology, medical entomology and sociology to facilitate analysis of the socio-ecological processes that encourage the proliferation of *Aedes aegypti* and *Aedes albopictus* mosquitoes, species that vector dengue, chikungunya and Zika virus. Research focused on individual homes with gardens, identified as particularly conducive to the life-cycle of so-called domestic mosquito species (Carrieri *et al.*, 2011; Scott *et al.*, 2012). Comparing mainland France and the French Antilles allowed us to examine the environmental and cultural similarities and differences in territories that, at first glance,

[52] The PROLITENSAN programme – "*Proliferation of land- and marine*-based *coastal species* with a strong effect on the environment and health: a comparison between mainland France (Mediterranean coast) and overseas France (coasts in the French Antilles)" – was financed by the Foundation de France. Focused on the issue of mosquitoes, it brought together a team of researchers and technicians from LPED at Aix-Marseille University (Cécilia Claeys, Christine Robles, Valérie Bertaudière-Montes, Magali Deschamps-Cottin, Hervé Tepongning Megnifo, Manon Sense, Pauline Bravet, Laura Weill, Claire Demerrisse, Hubert Mazurek, Louis Arrhegini), the Institut Pasteur in Guadeloupe (Christelle Dollin, Florence Fouque), the Mosquito and Vector Control Department of Guadeloupe (Joël Gustave) and Martinique (Manuel Étienne, Renélise Pélagie-Moutenda, Fabrice Sonor, André Yebakim) and Entente Interdépartementale de Démoustication de Méditerranée (Charles Jeannin).

appear very different in most respects other than their mutual belonging to the French nation.

This chapter will focus on the qualitative analysis of the discourse and practices of inhabitants.[53] We will begin by presenting the methodology and corpus of data. Then our analysis is divided into four sections. The first section identifies types of gardens and management approaches that encourage or discourage the presence of vector mosquitoes. We examine the different statuses of water for utility and/or aesthetic and leisure purposes. The second section analyses the ambiguous relationship that amateur gardeners have with pesticides. We describe what we have called "ladybug syndrome". The third section argues that the consumerist uniformization of gardens has encouraged a proliferation of vector mosquitoes. Emphasis is placed on something that resembles a culturally constructed "Diogenes syndrome". The fourth section addresses the delicate topic of the responsibility of actors in vector control (VC). It examines the multiple demands expressed by multifaceted public authorities and the difficulty in getting economic actors from the garden and habitat sectors on board. Lastly, we conclude by formulating recommendations for gardens and habitats that address both sustainability and vector control.

Methodology

To our knowledge, no previous research has ever combined medical entomology, plant ecology and sociology in the context of analysing the factors underpinning the proliferation of *Aedes albopictus* and *Aedes aegypti*. This first attempt showed the feasibility of the approach, its heuristic value and that it can be reproduced and transposed to other contexts. Within the context of the PROLITENSAN project, we focused on four research zones located in urban and suburban coastal areas: two in mainland France (Nice and Marseille) and two in the French Antilles (Petit-Bourg in Guadeloupe and Le Vauclin in Martinique) (*figure 1*).

---

[53] A presentation of the quantitative interdisciplinary analyses from this research project may be found in Claeys *et al.* (2016).

Figure 1. *Localisation of the studied areas.*

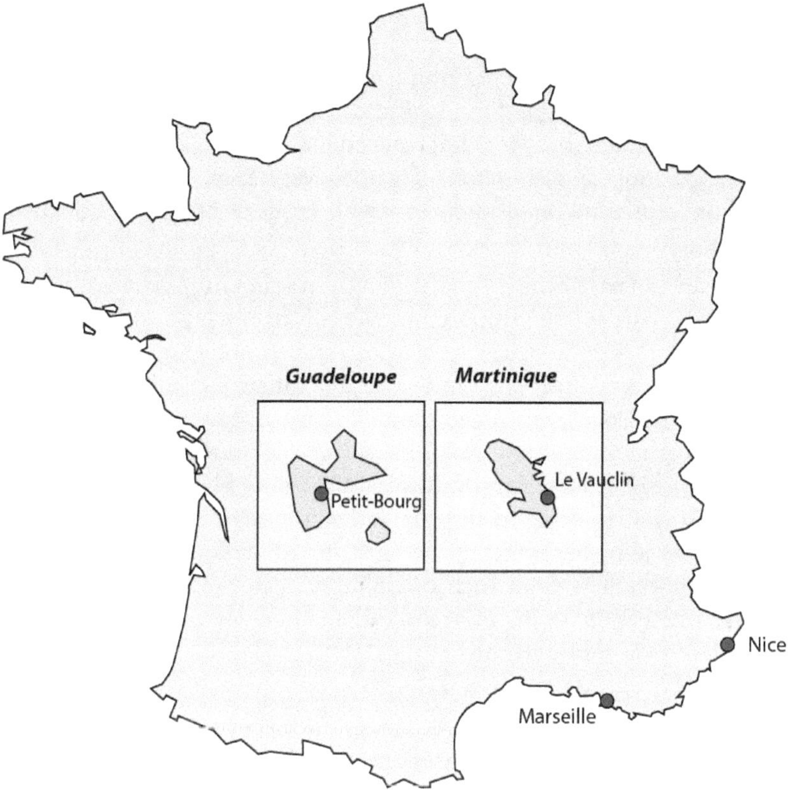

The two overseas municipalities were chosen for their similar regular climate and because the population in both zones has complained to mosquito control services and/or the municipality about mosquito-related problems, as well as for the regular presence of dengue fever cases. In southern mainland France, two field areas were studied, one in the city of Marseille and the other in the city of Nice. While in the Antilles the protocol covered the entire coastline of the selected municipalities, in mainland France it was necessary to divide the area due to the size of the two agglomerations. A transect was drawn that ran through both municipalities from the sea to their interior neighbourhoods. These transects ran through areas where the population had most expressed discomfort related to mosquitoes to mosquito control and/or municipal services. It was also necessary that the zones selected contain private homes with gardens. The study unit was the single-family home including its garden. The geo-referenced samples included 160 homes and their respective gardens, divided equally across the four sites. Data were simultaneously collected from each sample unit (i.e. single-family home). These included entomological data, with an inventory of larval breeding sites and indoor adult collections; botanical data including the composition and structure of garden vegetation; and sociological data from semi-structured interviews with heads of households or their partners.

In the context of this interdisciplinary research, the data collection phase was particularly important. We opted for the simultaneous collection of entomological, ecological and sociological data, which allowed us to test direct correlations between insects, plants and human behaviour. It also allowed for *in situ* reflexivity. Following the first phase of sociological interviews, researchers from all three disciplines and inhabitants discussed the entomological situation of homes and gardens, thus making it possible to directly gather the reaction of inhabitants. Finally, this joint fieldwork encouraged interdisciplinary emulation by teaching researchers from each discipline to better communicate with others, whether in terms of technical, methodological, theoretical or epistemological considerations.

The analyses presented here are based on qualitative data from 160 semi-structured interviews, field books (post-interview direct observation) and the recording of the April 2016 seminar in Fort-de-France (Martinique) to present the results of the PROLITENSAN programme which was attended by fifteen professionals and local actors from the non-profit sector involved in the design and management of gardens.

Amateur gardeners, inadvertent breeders of vector mosquitoes

Amateur gardeners can inadvertently breed mosquitoes. *Aedes aegypti* and *Aedes albopictus* seek out small sources of clean water in which to lay their eggs. They also look for shaded areas to rest. When seeking a resting zone, *Aedes aegypti* prefer the interior of homes whereas *Aedes albopictus* prefer the shaded greenery of gardens. Both these *Aedes* species feed on the nectar of plants and their females require a blood meal for their eggs to reach maturity. They lay their eggs in aquatic habitats in which they become larvae, then pupa before hatching into adult mosquitoes. Single-family homes with gardens offer both these *Aedes* species a living environment well adapted to their life cycle, with plants, water, shade and mammals to bite (Jansen and Beebe, 2010). Given this, certain aesthetic and cultural choices by gardeners can encourage or conversely reduce the formation of larval habitats, as well as rest areas for adult mosquitoes. Similarly, the lifestyle of inhabitants may also encourage or limit their exposure to the bites of these vector insects.[54]

Water and its different uses in gardens are key factors in explaining the positive presence of larval habitats.[55] Gardens with fountains, basins and pools[56] tend to be conducive to the presence of larval habitats, as well as those that engage in rainwater recovery. In the gardens visited, water had two main roles: for aesthetic and leisure purposes, or for utility and sanitation purposes. Water for aesthetic and leisure purposes is generally the product of a choice, whereas water for utility purposes is more of a necessity, but the boundary is nevertheless porous between choice and necessity in the sense that some uses of water may belong to both registers.

The reasoning behind rainwater recovery, for example, is very different in the French Antilles and mainland France (*Figure 2*). In mainland France, it is generally part of a well-intentioned middle- and upper-class effort to be environmentally friendly; the harvested rainwater is used primarily for watering gardens and potted plants. The inhabitants interviewed mentioned the virtues of rainwater compared to "treated" and "chlorinated" tap water they believed to be less good for their plants. Such inhabitants also tended to drink bottled mineral water themselves. The economic side of rainwater recovery was either secondary or explicitly denied by the most affluent inhabitants interviewed.

---

54 This was how Soulancé *et al.* (2011) were able to observe during the Chikungunya epidemic in Reunion Island that inhabitants who spent more time in their gardens had higher prevalence rates.

55 A potential larval habitat is a container with the adequate properties to host *Aedes Aegypti* or *Aedes albopictus* larvae, but which does not contain any larvae at the time of inspection. A positive larval habitat contains the larvae of *Aedes aegypti* or *Aedes albopictus*.

56 Pools do not automatically constitute a larval habitat unless they are not maintained. Larval habitats are generally found in the installations and objects that are directly next to or associated with them: skimmers, patios, showers, garden furniture, parasol bases, children's toys, etc.

Figure 2. *Rainwater recovery practices (photo credit: Claeys).*

2.a. An inhabitant in Guadeloupe (Antilles) emptying a bucket of recovered rainwater at the request of an ARS field agent

2.b. An inhabitant in the Nice region (mainland France) emptying a rainwater recovery basin at the request of an EID field agent

Rainwater recovery is very widespread in the Antilles (Mieulet, 2015) where it is part of the native American and Creole heritage. While the principle has been handed down from generation to generation, some know-how appears to have been lost with changes in the types of recipients used. In the past, earthenware jars were carefully buried whereas today uncovered barrels and tubs are used, recycled from elsewhere or purchased at retail outlets. Vector-control (VC) awareness-raising campaigns specifically target such practices (*cf. intra*) and recommend the permanent and complete covering of recipients. And yet field observation has shown how little inhabitants follow such instructions. Firstly, many people do not consider water that is used regularly to be stagnant. Moreover, copper (via the immersion of coins) or rust (via the use of metal barrels) are often thought to be larvicidal, much like putting in a few drops of bleach.[57] Inhabitants who do cover their rainwater recovery barrels do not seal them entirely or systematically, and often underscore the hassle involved in doing so. Rainwater recovery in the Antilles has a distinctly economic and technical character. In both Martinique and Guadeloupe, the water distribution networks are chronically defective. Unexpected water cut offs are common and may last several days, in addition to planned cut offs tied to maintenance on an obsolete network regularly damaged by an array of natural tropical hazards. Recovered rainwater may be used by inhabitants to water their gardens, although it is sometimes also used for cleaning and doing laundry. In poorer households, it may also be used for bathing, and even for cooking. With or without water cut offs, tap water in Martinique is the most expensive anywhere in France (D'Ornellas 2010), which further encourages rainwater harvesting.

Regarding sanitation water, there was a marked difference in our study sample between the Antilles and mainland France. While 92% of the homes visited in mainland France were connected to the sewer system, this was the case for only 25% of homes in the Antilles – and those not connected to the network in the Antilles were more likely than others to contain positive larval habitats (58% had larval habitats versus 42% of the others) (Claeys *et al.*, 2016). On this topic, the same analysis was not possible in mainland France as it would have required more homes in the sample that were not connected to the public sanitation network which is relatively uncommon within urban zones in Marseille and Nice. If we consider only the situation in the Antilles, it appears that not being connected to the public sanitation network encourages the presence of larval habitats. And yet nearly 60%

[57] Copper indeed has larvicidal properties (Bellini *et al.*, 1998) but it needs to be present in large enough quantities.

of homes in the Antilles are not connected to this network (Adin, 2018), again underscoring that these overseas French territories are confronted with precarious infrastructure and technical and economic inequalities that resemble those facing the global south.

The analyses conducted in plant ecology demonstrated clear floristic differences between mainland France and the Antilles which corresponded to their distinct bio-geographical zones. Despite this, there were similar processes at play in both. Some research has shown the individual potential of certain plant species to attract or repel mosquitoes (Samson *et al.*, 2013; Reiskind *et al.*, 2010). Despite the presence of some of these (e.g., *Cymbopogon citratus, Pelargonium citronnellum*), our study did not show any significant correlation between the global floristics of a garden and the presence or absence of vector mosquitoes in the observed sample. It is possible that this lack of significant correlation was because these attractive or repellent species were present only in limited quantity in gardens. The structure of gardens did play a clear role, however: on both sides of the Atlantic, open gardens[58] were less likely to house vector mosquitoes than shaded gardens.[59]

This open/shaded effect can be explained by several factors. The strictly structural effect is tied to the role of plant cover in the life cycle of *Aedes albopictus* and *aegypti* mosquitoes. Shrub vegetation arranged in hedges or beds provides mosquitoes with excellent resting zones (shade and dampness), in addition to being a feeding area. Two situations encountered during our fieldwork were particularly telling. An inhabitant interviewed in Guadeloupe told the team of researchers that he had planted a bed of citronella plants around his patio in hopes of repelling mosquitoes, but that the opposite had occurred. Citronella is traditionally used as an essential oil or burned[60] to repel mosquitoes. Yet the citronella bushes actually provided mosquitoes with a resting zone and the watering of this planted bed further provided favourably damp conditions. In this example, the bed of citronella bushes was planted near a patio upon which users provided a blood meal for female mosquitoes. In mainland France, several inhabitants planted scented geraniums, advertised by the garden centres that sell them for their mosquito-repellent capacities. None of the inhabitants interviewed noticed a significant decrease in

58 Open garden: with little shade.

59 Shaded garden: gardens with heavy tree cover.

60 The so-called "smoking" technique is very common in the Antilles.

mosquito presence after having planted the geraniums. In Nice, one inhabitant who planted an entire bed of scented geraniums across his whole garden mentioned trying to lay on a chaise longue in the middle of this plantation without any success in terms of repelling mosquitoes.

In/desirable nature: the dilemmas and paradoxes of gardeners vis-à-vis pesticides

Unlike many of the alien exotic species that interest ecologists (Walther *et al.*, 2009), *Aedes aegypti* and *Aedes albopictus* did not escape from gardens to disrupt the natural balance of endangered ecosystems. In fact, the opposite is true: they are attracted to gardens. Although classified as domestic, these mosquitoes have not been domesticated. They actually embody the undesired and undesirable intrusion of a wild species that has abandoned the nobleness of nature to embrace the soiled stain of urbanity. These domestic mosquitoes are perceived as particularly bothersome by inhabitants since they invade the personal sphere of homes and private gardens. Their presence tends to be experienced as an impurity in the anthropological sense of the term – i.e., an intrusion that is both physical and symbolic (Douglas, 1966).

The amateur gardeners we met described what we will call the "paradox of the ladybug". During our fieldwork, the sociological interviews first invited inhabitants to talk about their gardens in general, how they garden and their use of inputs, before addressing the issue of vector mosquitoes. All the inhabitants we met in mainland France claimed to engage in different types of environmentally-friendly gardening. Ladybugs were mentioned repeatedly by inhabitants as a symbol of pleasant and beneficial nature in the city. Referred to in French as "Godly creatures" ("*bêtes à bon dieu*"),[61] they belong in the register of imagination, childhood stories and are considered to bring good luck. The inhabitants interviewed wished for them, since they were associated with the principal of integrated control. These pleasant insects are indeed a predator to aphids, the bane of any gardener's existence. All the inhabitants interviewed in mainland France felt it was good to practice environmentally-friendly gardening. And when

[61] The expression is thought to derive from a legend with several different versions but which share the common idea that during the Middle Ages the presence of a ladybug interrupted the execution of a man sentenced to death for a crime he had not committed.

they admitted to using phytosanitary products, it was often given as an excuse (the parasite resisted organic treatments, the invasion of said plant or animal species was getting out of control, etc.) or minimized (mentioned very localized and/or very exceptional and/or very lightly dosed usage, etc.). This ambiguous relationship with phytosanitary products is quite common in Western gardening and has been particularly well documented in the seminal research of Paul Robbins (2007) regarding the USA. In the Antilles, the wealthier people met tended to have an attitude similar to that observed in mainland France. Poorer inhabitants however – who, in our sample, were all Creoles[62] – more commonly drew on the register of know-how passed down from one generation to another and/or the too great expense of phytosanitary products available in stores. In both cases, they tended to partake in what Joan Martinez-Allier (2014) refers to as the environmentalism of the poor.

While the gap between discourse and practice in terms of environmentally-friendly gardening seemed less stark in the Antilles than in mainland France, the situation was more complex regarding mosquito control. The enmity of inhabitants on both sides of the Atlantic towards mosquitoes trumped their love for ladybugs in particular and for nature more generally. There is no doubt: not all animal figures are welcome in the domestic sphere (Blanc 2000). The real or stated moderation of environmentally-friendly gardeners in their use of chemical inputs indeed gave way to calls for large-scale insecticide spraying in the name of controlling nuisance and vector mosquitoes. The inhabitants we met were ambiguous in their discourse on the use of insecticides for mosquito control. Our previous research in mainland France (cf. *infra*) has shown that the populations interviewed mention the potentially harmful effects on health and the environment of large-scale mosquito control via the spraying of insecticides, while also calling for its implementation. Although they claim not to be overly concerned about the emergence of local arbovirus epidemics, they justify calls for mosquito control by citing health-related arguments, considered to be more acceptable than the mere desire for personal comfort. The interviews conducted in the context of PROLITENSAN further confirmed this tendency. They also highlighted a certain ignorance among the population encountered in mainland France regarding the different mosquito control techniques.

[62] French law limits or prohibits the compilation of ethno-racial statistics, but it is difficult to ignore that in the Antilles, like the in USA, people's income level tends to be inversely proportional to the darkness of their skin.

Some people did not hesitate to call for their neighbourhood to be flown over by helicopters spraying insecticides in order to rid them of this "new mosquito" and its particularly urticant habits. In the meantime, they relied on an arsenal of store-bought insecticides: electronic repellents, incense coils, body sprays, etc. The manufacturers and distributors of insecticides and repellents, neighbourhood pharmacists and large retail stores very quickly seized on this new business opportunity. During the summer season, pharmacists display soothing creams and repellents in their window displays, particularly the special "tropics" range. In large retail chains, the packaging on electronic mosquito repellent boxes now contains the words "tiger mosquito". The accounts collected from pharmacists interviewed in the Alpes Maritimes and Marseille corroborate this (Claeys and Mieulet, 2013). The latter confirmed – albeit always off the record – that the arrival of tiger mosquitoes in the region was good for business.

The situation is more complex in the French Antilles where there is nothing new about the presence of *Aedes aegypti*. What is new, however, is the evolution of mosquito control policies. The genetic mutation of *Aedes aegypti* in the Antilles and the fact that it is now capable of resisting insecticides encouraged mosquito control and VC services to limit the use of insecticide spraying (cf. *intra*). Spraying is now mainly limited to health crises and the localized targeting of epidemic outbreaks. And yet the inhabitants we met in the Antilles spoke with nostalgia about a recent past in which mosquito control services conducted insecticide treatments more regularly and on a larger scale. These inhabitants repeatedly conjured memories of trucks projecting a mist of insecticide in the streets. Yet their discourse was paradoxical. They called for regular mosquito control via spray trucks while also raising doubts about its efficiency. Some inhabitants had heard about the resistance of *Aedes aegypti* to insecticides while others had noticed it first-hand. Rather than a blatant contradiction, this was more of a "last resort" solution. The use of spray trucks was considered "better than nothing" and, even if their efficiency was not convincing, the underlying idea was that a few less mosquitoes was still an improvement. The poorest households in the Antilles also underscored the high cost of individual protection measures (sprays, lotions, diffusers, etc.). Further, all those with whom we spoke in the Antilles mentioned the declining efficiency of these individual protection measures against mosquitoes. Electronic repellents and indoor sprays were listed as the least effective – and yet people continued to use them for lack of a better solution, once again to the great satisfaction of shop owners. None of these different mosquito repellents actually reduce the presence of mosquitoes at the source, however; rather, they are a temporary (and costly) barrier that constantly needs to be re-erected between an anthropophilic insect and the human skin.

The standardization of gardens and consumerism

Research in the social sciences and humanities has clearly shown the socio-economic and cultural processes at play in the fabrication of gardens. Their closed nature and status as a domestic environment have tended to transcend periods and cultures (Dubost, 1997; Menozzi, 2007; Larbey, 2013). Their aesthetics are a reflection of their different uses – and these uses are situated within a physical and social time and space. For centuries, ornamental gardens were reserved for the wealthy. For most people, gardens were long a food source above all. In mainland France, the economic boom following the Second World War encouraged the development of single family housing. There was a relative democratization of this type of housing with ornamental gardens (Frileux, 2010), although in the regions of mainland France most subject to real estate pressure, such as along the Mediterranean coast, single family homes with gardens nevertheless remain accessible mainly to the middle and upper social echelons. In the Antilles, on the other hand, single family housing has always been predominant. Today, the range of housing is very broad, extending from plantation homes with their ornamental gardens to hovels with miniscule yards – and an entire gamut of dwellings inspired to varying degrees of precision by the traditional island hut model and its Creole garden. Creole gardens are "small, more or less well defined spatial units of agricultural production located next to single family homes, very common in rural and urban areas in the Antilles. Focused primarily on self-consumption, they are called 'Creole' because they are characteristic of a cultural and agricultural way of living inherited from the precolonial and colonial periods" (Marc, 2011, translated here). Creole gardens never disappeared and even experienced a renewal in popularity following the social movements of 2009[63] which led to the creation of local farmers' markets aimed at reducing the dependence of the local population on large shopping chains selling expensive products imported from mainland

[63] Social movements (general strike and demonstrations) which led to the emergence of LKP (Liyanna Kont Pwofitasyon), a collective of trade unions and associations.

France (Bonniol, 2011). When space permits, Creole gardens also sometimes have ornamental ambitions (Benoît, 2002).

In both mainland France and the Antilles, the population is truly attracted to gardening (Larbey, 2013; Marc, 2011). The success of gardening magazines and large horticultural distributors are proof of the current appeal for gardening (Dubost, 1994). They help shape the aesthetic canons of contemporary gardens (Marco *et al.*, 2010) and are part of the classic process of socially fabricating taste, creating tension between the efforts of some to be unique and the strategies of others to imitate.[64] The same popular gardening magazines are distributed in mainland and overseas France. A comprehensive overview of the main stationery shops and bookshops in Fort-de-France and Point-à-Pitre in April 2016 allowed us to confirm that these establishments sold only gardening magazines edited in and for mainland France. Similarly, a few large horticultural chains are predominant in the mainstream distribution channels in both mainland France and the French Antilles. The plants sold are primarily products from the international horticultural trade in which Dutch greenhouse production is central (Dubost, 1994).

This cultural and commercial context encourages a standardization of gardening models (Benoît, 2002; Frileux, 2010). Ornamental basins and fountains, potted plants, saucers and cachepots are all inspired by the same Western aesthetic and promoted by the large garden centre chains and specialized magazines. Similarly, from the most luxurious infinity pools to small inflatable pools filled with runoff water, people on both sides of the Atlantic share the same attraction for aquatic games at their homes, adapted to their financial means. During the seminar to present the results of the PROLITENSAN programme in Fort-de-France, a workshop was organized in partnership with local institutional actors to exchange with the non-profit sector and independent professionals involved in the landscaping and garden sectors in Martinique. The professionals underscored the loss of traditional Creole knowledge and know-how with respect to gardening, including in terms of vector control. One person specialized in Creole and sustainable gardening made the following comments during the meeting: *"I spend my time fighting against preconceived ideas imported from elsewhere. That's what we can call it. And if mosquitoes cause so many problems* [...] *it is in large part because there are external models that are copy-pasted and do not at all correspond with what should be done to do things right. For example, I*

64 According to the analytical framework proposed by Pierre Bourdieu.

advocate for what is called lasagne-bed gardening, which is a technique that lets you garden using very little water, for example. So it does not encourage the infamous vector that we're talking about today. And people tell me they're surprised and didn't know such a thing existed..."

In mainland France, the Mediterranean garden has entered the aesthetic canon disseminated by magazines and garden centres. While for botanists the term "Mediterranean garden" refers to a specific list of local species, for horticultural professionals and their clients – who easily deviate from the laws of botany – it is synonymous with a stereotyped vision thought to sell well (Claeys, 2010). And even if Arab-Andalusian gardens are often mentioned as a source of inspiration for contemporary aesthetic canons, this is forgetting that the latter – a privilege of the wealthiest – were indeed organized around water and its mastery, but of water that circulated in a complex network that was managed by a dedicated workforce (El Faïz, 2016). Along the Mediterranean coast, most of the plants introduced in gardens for their aestheticism and the image they create are exotic species (Marco *et al.*, 2010). Given that most of these are not adapted to the dry summer season, they require a lot of watering, often combined with the use of saucers to help conserve water. Aquatic installations are also common (fountains, basins, pools, etc.) and sometimes lead to the production of larval habitats when they are neglected (poorly maintained) or when plant species likely to retain water are introduced (water cabbage, water hyacinths, etc.).

The mosquito and VC control department in Guadeloupe was one of the first to attempt to raise awareness among garden centre professionals. Its director told us how, during a particularly virulent dengue epidemic, he managed with a great deal of effort to convince garden centres to stop selling saucers, which are particularly conducive to the formation of larval habitats. As soon as the epidemic was over however, the shops put the saucers back on their shelves.

Another factor that encourages the formation of larval habitats is overcluttered gardens (Soulancé *et al.*, 2011; Faye, 2017). This phenomenon was visible at homes visited in both mainland France and the Antilles. The overcluttering of gardens is common mainly among the poor and/or elderly. It is sometimes akin to Diogenes syndrome,[65]

---

[65] "Diogenes syndrome is a behaviour disorder involving a tendency to accumulate objects (hoarding disorder), neglected personal and domestic hygiene and, very often, marked social isolation without any desire to complain about this situation.

although the situations observed in the field during our interviews were not solely the result of this psychiatric condition. They were also a product of social context. For some elderly people, their garden is a resting place for objects accumulated throughout their lives, which they hold on to for possible re-use, out of nostalgia or because they are no longer physically strong enough to move or dispose of them. For the poorest, each object is conserved with its possible re-use in mind or due to the inability to remove it for technical or financial reasons. When inhabitants are both old and poor, the two processes can be compounded. This type of situation was particularly common among the houses visited in the Antilles. There were several factors at play. To begin, the share of poor households within our sample was larger in the Antilles than in mainland France, revealing a relegated coastline in the Antilles and a socially discriminating one along the Mediterranean (Claeys, 2017). Second, compared to mainland France, there is a public service deficit in the Antilles in terms of the removal and treatment of waste. The case of Vehicles Off Road (VOR) is symptomatic of this. The poor development of public transportation, its high cost, irregular schedules and poor territorial coverage has long encouraged the proliferation of cars in the French Antilles. Moreover, low income and high unemployment rates (Michel and Theulière, 2010) have further encouraged the development of used vehicle and recovered spare parts markets. According to existing regulations, the removal of VORs in the Antilles is free of charge and must be undertaken by a certified professional – but the free nature of the service applies only if the VOR is complete. Moreover, professionals have the right to charge a towing fee. Given this, the stockpiling of automobile hulks and spare parts in people's gardens is commonplace. Yet VORs are conducive to the formation of larval habitats. During endemic periods, voluntary policies are implemented to remove VORs abandoned on public ways, but such initiatives are not ongoing.

Standardization, loss of local know-how and knowledge, alienating consumerism and poor-development are all factors that work to aggravate the proliferation of domestic vector mosquitoes. The causes behind such proliferation and the subsequent epidemic risk are multifactorial. They

---

Combined, these conditions lead to an insalubrious life and dilapidation of the home in which the person with Diogenes syndrome lives. In a large majority of cases, the syndrome develops following a psychological shock such as the death of a loved one or a radical change in situation. It mainly affects the elderly" (www. syndrome-diogene.fr, translated here).

are based on socio-economic, cultural and political processes and they challenge individual and collective, public and private responsibility.

Behavioural habitats versus structural habitats: the difficult task of sharing responsibilities

The field observations conducted in the context of the PROLITENSAN programme, as well as previous research have allowed us to identify two main types of larval habitats that we will call *behavioural habitats* and *structural habitats* (Claeys and Mieulet, 2013; Mieulet, 2015; Mieulet and Claeys, 2016).

The presence of behavioural habitats arises when inhabitants fail to apply the VC measures recommended in public awareness-raising campaigns.[66] This failure to follow the recommendations of information campaigns may be due to ignorance about or poor understanding of VC measures, or based on a refusal to implement them. In the Antilles, situations of ignorance are rare (Setbon and Raude, 2008), since the population is the target of many highly visible awareness-raising efforts. In mainland France, inhabitants in the regions recently colonized by *Aedes albopictus* are sometimes unaware of the VC measures that should be adopted. Our longitudinal research (Mieulet and Claeys, 2016) nevertheless indicates regular improvement in the knowledge level of the population along the Mediterranean coast. The existence of a multitude of domestic larval habitats in recipients containing clean water challenges typical Western vernacular taxonomies that associate mosquitoes with natural spaces (wetlands) and dirty places (septic tanks, sewers, dirty water). In mainland France, where the introduction of *Aedes albopictus* is recent, this situation is experienced personally and socially as a potentially stigmatizing blemish. In the Antilles, on the other hand, the population is more aware of the domestic nature of *Aedes aegypti* and it tends to be experienced less as a blemish than in mainland France. The proliferation of the insect is mainly associated with the tropical nature of the local climate, yet without entirely shedding its intruder status. Here or elsewhere, mosquitoes are triply seen as "spatially intrusive animals" (Mounet 2012) – from the nocturnal intrusion of their pesky buzzing, to the cutaneous intrusion of their irritating bites, to the viral intrusion of potentially deadly diseases.

66 These campaigns call for the monitoring and destruction of larval habitats and notably insist on saucers, rainwater recovery barrels, gutters and bulky items stored in gardens.

This rejection causes responsibility to be placed on others and particularly on public authorities. The shifting of responsibility from inhabitants towards public authorities is part of a broader refusal to accept the decline of the welfare state. The underlying reasons differ however on either side of the Atlantic and this difference can be quite marked. In Nice, for example, one inhabitant who was very bothered by the presence of *Aedes albopictus* felt that it was up to public authorities to solve the problem since she was a taxpayer – she even pointed out that she was subject to the wealth tax. Overseas, the legacy of colonialism and slavery, geographical distance from the central state power and economic weaknesses fuel a sense of abandonment (Affergan, 2006). Given this, requests to public authorities for mosquito control solicit a welfare state deemed too far and not sufficiently attentive to the unique specificities of the Antilles.

Regarding structural habitats, their presence is directly tied to architecture, landscaping and urban forms. Such habitats can be encouraged when existing construction norms are not respected (gradients, evacuation systems, etc.), but also by certain architectural and/or landscaping choices, and by some urban forms and their (non) management. Like the "bottomless barrel of the Daughters of Danaus", such larval habitats are perpetually (re) filled, eventually wearing down the good will even of the best-informed inhabitants. The invisibility and/or inaccessibility of many *Aedes albopictus* and *Aedes aegypti* larval habitats makes their presence unsuspected and helps them escape the watch of even the most willing inhabitants. This ergonomic barrier highlights the structural role of buildings and gardens in the formation of larval habitats, but also of urban sanitation networks (particularly when they are malfunctioning or absent).

From landscapers to box store garden centres, architects to hardware stores, without forgetting the "trendsetting" gardening and decoration magazines – all these garden and construction sector professionals can encourage the presence of larval habitats through their technical and aesthetic choices. Awareness-raising campaigns target mainly the general public – i.e., the last link in the chain of technical and aesthetic decision-making – thus placing responsibility on inhabitants that should be shared by the large array of economic actors who contribute to the creation of potential larval habitats.

Existing regulations in France regarding construction norms do not yet explicitly address the prevention of structural larval habitats. If carefully respected, some of these norms could nevertheless already contribute to vector control. That is the case with the building code (*Documents*

Techniques Unifiés – DTU) related to the technical management of rainwater in built infrastructure: roofs, terrace-roofs and raised patios, as well as gutters, drainpipes, valleys and downspouts are the focus of a series of norms related to both their shape and installation. But while the norms established in the DTU share with VC principles the desire to avoid the creation of zones of stagnant water, many professionals in the construction sector tend to show a certain degree of freedom and approximation vis-à-vis these technical norms (Cardi, 2017). Indeed, regulation in the construction sector in France is particularly burdensome and complex, and is often perceived by professionals more as a constraint than a solution. Even when scrupulously respected, these technical norms do not make it possible to avoid regular maintenance of rainwater evacuation devices and of built infrastructure more generally (e.g., the inspection, cleaning and replacing of damaged or worn out parts).

Since 1996, in addition to the norms that exist under French law, there are fourteen targets within the High Quality Environmental (HQE) standard, replaced in 2015 by the HQE Reference Framework. Very widespread in contemporary construction, this approach is concerned above all with saving energy and raw materials. The topic of vector control is not explicitly addressed. Among the practices recommended in the HQE approach, some comply with the principles of VC (e.g., waste management and the ongoing monitoring of performance), while others go against it – particularly those pertaining to rainwater recovery, its storage and use. The new HQE Reference Framework proposes a global and contextualized vision of construction among the five principles that compose it, which include concern for the health of human inhabitants and for the biodiversity present in and around buildings. With such renewal, it is possible to imagine the development of new building techniques that are more sustainable and also attuned to vector control.

Finally, the multitude of messages addressed to the population by the multifaceted public authorities tends to affect people's understanding of VC techniques. Once again, the issue of rainwater recovery provides a good example. In the context of developing sustainable housing, the institutions in charge of environmental policies recommend rainwater harvesting and use (Souami, 2012), an approach that can be conducive to the formation of larval habitats. The Ministry of Ecology, sustainable development and energy has a set of guidelines for rainwater harvesting and use entitled "Rules and Good Practices for Installers". The approach is environmental and aims "to fulfil one of the commitments of the Grenelle Environnement".

Its arguments are incentive-based, with information about "tax credits" and potential "financial aid", and its advice is technical regarding the "design, measurement, implementation, installation, maintenance and upkeep" of rainwater harvesting and recovery equipment. While health concerns are not ignored, they are limited to the non-suitability of rainwater for drinking. The document notably warns against "health risks tied to the coexistence of a rainwater network and the public distribution network." The technical instructions recommend installing "mosquito covers" on aeration vents in the mechanism and more broadly recommend the installation of a "permanent solution to avoid insects and small animals entering the tank" to limit the risk of polluting the water. There is no mention of vector issues or the monitoring of larval habitats, however. In addition to such ministerial messages, professionals, distributors and installers of rainwater harvesting equipment also have their own communication strategies. An internet search conducted on 4 August 2016 on the websites of the leading French distributors and installers did not identify any information related explicitly to the prevention of larval habitats forming in rainwater storage tanks. Thus, in terms of water management at private homes, the government or, more precisely, public authorities appear to send contradictory messages to the population. The multifaceted nature of public decision-making has been repeatedly pinpointed. The Ministry of Ecology is particularly exposed to such tension between its environmental prerogatives and those of the ministries in charge of different and potentially antithetical issues (Aspe and Jacqué, 2012). And yet here, the contradiction is based less on fundamental discord than on a lack of effort to coordinate messages. VC does not automatically call for the abolishment of rainwater harvesting and use, but rather for its inclusion in technical prevention approaches to combat the formation of larval habitats.

Recommendations: For a habitat and urbanism that are both sustainable AND anti-vector

The difficult choice between protecting the environment and preventing arboviruses leads to multiple orders from a multifaceted public authority and to contradictory desires expressed by a polymorphous population. The situation is more complicated than an opposition between protagonists with different interests since it also highlights tension within each group of actors, as well as individual dilemmas. Gardeners dream of an ecological garden filled with ladybugs, but without mosquitoes; public authorities advocate for sustainable housing and rainwater recovery while calling on inhabitants to empty domestic larval habitats to reduce epidemic risks.

Opposing nature protection and the protection of human health is a technical and ethical impasse inherited from modernity. There are other options. The recommendation by our interdisciplinary team to envisage a ***habitat and urbanism that are sustainable and anti-vector*** could be a first step, among others. The results of our research confirm the previously identified need (ACSES, 2006; Hounkpe, 2012) to better include the prevention of larval habitat formation right from the design phase of buildings, as well as including it in the drafting of urbanism projects. To date, French public policies have tended to handle environmental and health issues in a disconnected manner, with a splitting of roles between the ministries of the environment and health. The result is that architectural and urbanism recommendations formulated in the name of sustainable housing tend to encourage the proliferation of vector mosquitoes while, conversely, certain vector control strategies are not without negative ecological and eco-toxicological consequences.

Given this, our team recommends defining, institutionalizing and promoting *habitats and gardens that are sustainable AND anti-vector*. A first set of recommendations related to gardens can be formulated based on our interdisciplinary observations in ecology, entomology and sociology.

1) In the Mediterranean region, it is preferable to favour local species. They are adapted to the climate and soil, require less watering and can be planted in the open ground rather than in potted containers which, given the presence of saucers underneath, constitute potential habitats. The use of mulch on plant beds is another way to limit the loss of water through evaporation and to reduce watering, thus reducing the dampness of the environment, as well as the presence of residual puddles that can be potential larval habitats.
2) In the Mediterranean region and the Antilles, open areas with discontinuous plant stratification should be encouraged in order to avoid the creation of wet shaded zones which constitute rest areas for adult mosquitoes. This recommendation is not necessarily contradictory with the presence of trees in gardens to provide shade for inhabitants or food crops in Creole gardens. It is simply necessary to not create continuous cover and thus to prune trees in such a manner as to ensure that their crowns are not too low or connected, and also to avoid planting shrubs and bushes under them.

3) In a similar vein, the planting of trees whose crowns are above the roofs of houses should be avoided since fallen leaves obstruct gutters and spouts, forming unintentional water reservoirs (Gustave *et al.*, 2012).

These recommendations make it possible to envisage gardens that are less conducive to the presence of mosquitoes – i.e. *anti-vector gardens* – which in many respects also comply with planning and management recommendations for *sustainable gardens* (notably native species and water savings in mainland France).

Figure 3. *Structural habitats favoring mosquitoes reproduction.*

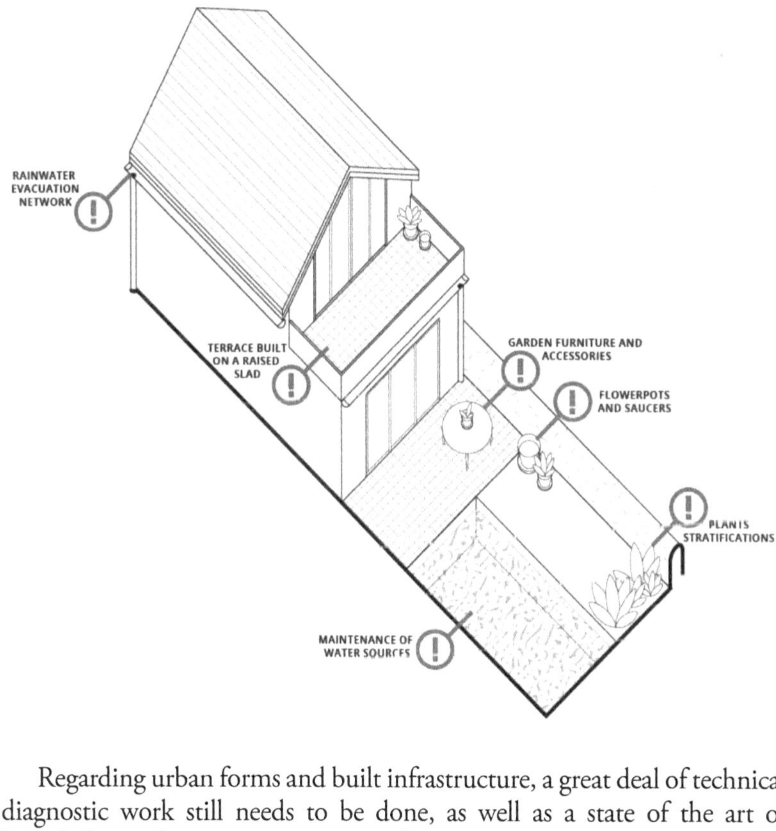

Regarding urban forms and built infrastructure, a great deal of technical diagnostic work still needs to be done, as well as a state of the art of knowledge and practices among professionals in the construction sector including all the building trades – architects, urban planners, masons, roofers and many others (*Figure 3*). This project is only just beginning.

To date, the alternative means of vector control explored in urbanism and architecture mainly recommend palliative solutions in the form of furniture to be installed in homes or public spaces. That is the case, for example, with the solar and wind-powered outdoor lighting system developed by a team coordinated by Dr Chong Wen Tong[67] (ResearchSEA, 2015) in the mechanical engineering department of the Faculty of Engineering at the University of Malaysia. To curb the risk of dengue fever in Malaysia, researchers included in the device a trap that attracts mosquitoes. It emits ultraviolet waves which, when they interact with certain parts of the device covered in titanium dioxide, produce the same CO^2 exhaled by humans, and draw the insects to the traps. In a similar vein, the French company *Techno Bam* (now called *Qista*) developed mosquito traps to address municipal needs. The traps are wooden boxes that attract and trap mosquitoes within a sixty-metre radius. The traps were tested in Sambuc, a hamlet near the Camargue natural regional park, where they decreased the nuisance caused by the presence of mosquitoes by 74 to 98% (Poulin *et al.*, 2016).

On a larger scale, the Jade Eco Park – an urban park project in Taiwan designed by architects Ricky Liu Associates and Philippe Rahm[68] with the "Mosbach" landscaping firm – has planned "*anti-mosquito*" areas. The different parts of the park were conceived in such a way as to create specific ambiences by using the pre-existing characteristics of the terrain (temperature, dampness and air pollution). Zones with both low pollution levels and few mosquitoes were equipped with an ultrasonic emitter capable of repelling insects, thus creating areas with little risk of insect bites for the most vulnerable populations (e.g., children and the elderly). While tools such as "anti-mosquito" lamps are largely anti-vector accessories, the very contextualized strategy implemented in Taiwan – based on the uses and natural features of a site – represents a more integrative approach.

Finally, to conclude this non-exhaustive list, we can also mention the Mosquito Contamination Device Project (MCD-Project) (McCaw, 2014), a project financed by the European Community's Seventh Framework Programme whose goal is to create economically viable anti-vector solutions for the poorest regions in the world. One of the devices created by the organization and called "Eaves Tubes" involves the insertion of simple tubes in traditional mud walls. In addition to fitting

---

67 ttp://www.researchsea.com/html/article.php/aid/9124/cid/2/research/smart_outdoor_lighting_system_with_mosquito_trapping_feature_powered_by_wind-solar_hybrid_energy.html

68 http://www.philipperahm.com/data/projects/taiwan/

perfectly into the construction of walls, they also provide an efficient aeration system for homes in tropical environments. The tube channels the odour of humans and the CO^2 produced from their breathing and uses them to attract mosquitoes which find themselves trapped in nets affixed to the end of the tubes. This solution nevertheless does not address the issue of mosquito bites outside the home.

LPED (Laboratory Population Environment and Development) continues to pursue its research into sustainable and anti-vector homes and habitats. A thesis drawing on architecture, urbanism and sociology has just been undertaken by Julie Cardi. This research aims to conduct diagnoses and formulate recommendations for buildings that address the double demand of sustainability and vector control. The endeavour is as such ongoing.

These research trends are based on an ecocentric ethic which considers it up to human beings to change their social organization and urban structures if they want to reduce vector-borne disease risk. Some might consider such an approach obsolete in the age of genetic engineering, which aims instead to eradicate mosquitoes or at least eliminate their vector competence through biogenetic modification. Only time will tell. But what we can already be sure of is the controversy surrounding the field (see Boëte, infra).

References

ACSES (2006). *Les gîtes larvaires des moustiques Aedes aegypti dans le bâti de Guadeloupe*, Rapport d'étude pour la DSDS de Guadeloupe.

Adin M. (2018). « Introduction, Assises Antilles-Guyane de l'assainissement non collectif, 27-28 juin, Palais des Congrès, Madiana, Martinique ». Online: www.eaumartinique.fr. Accessed 30 September 2018.

Affergan F. (2006). *Martinique : les identités remarquables. Anthropologie d'un terrain revisité*, Paris, PUF.

Aspe C. and Jacqué M. (2012). *Environnement et société. Une analyse sociologique de la question environnementale*, Paris-Versailles, MSH-Quae.

Bellini R., Carrieri M., Bacchi M., Fonti P. and Celli G.J. (1998). "Possible utilization of metallic copper to inhibit Aedes Albopictus (Skuse) larval development", *Am Mosq Control Assoc*, 14(4), 451-456.

Benoît C. (2002). *Corps, jardins, mémoires. Anthropologie du corps et de l'espace à la Guadeloupe*, Paris, Éditions CNRS.

Blanc N. (2000). *Les animaux et la ville*, Paris, Odile Jacob.

Bonniol J.-L. (2011). « Trois mois de lutte en Guadeloupe », *Les Temps modernes*, 662-663, 82-113.

Cardi J. (2017). *Architectures et villes du moustique : ce que l'expansion d'Aedes albopictus et ses conséquences racontent des nouveaux modes d'habiter humains*. Mémoire de master, ENSA Marseille.

Claeys C. and Mieulet E. (2013). "The spread of Asian tiger mosquitoes and related health risks along the French Riviera: An analysis of reactions and concerns amongst the local population", *International Review of Social Research*, 2(3), 151-173.

Claeys C. (2010). « Les " bonnes " et les " mauvaises " proliférantes : controverses camarguaises », *Études rurales*, 185, 101-117.

Claeys C. and Sirost O. (2010). « Proliférantes Natures : Introduction », *Études rurales*, 185, 9-21.

Claeys C., Robles C., Bertaudière-Montes V., Deschamps-Cottin M., Megnifo H.T., Pélagie-Moutenda R., Jeannin C., Sonor F., Dollin C., Sense M., Bravet P., Weill L., Demerrisse C., Mazurek H., Arrhegini L., Étienne M., Yebakima A., Gustave J. and Fouque F. (2016). "Socioecological factors contributing to the exposure of human populations to mosquito bites that transmit dengue fever, chikungunya and zika viruses: a comparison between mainland France and the French Antilles", *Environ Risque Sante*, 15(4), 1-8.

D'Ornellas C. *et al.* (2010). *Audit sur l'eau en Martinique*, Rapport n° 007091-01.

Douglas M. (1966). *De la souillure.* Paris, Maspero.

Dubost F. (1994). *Vert patrimoine.* Paris, Maison des Sciences de l'Homme.

Dubost F. (1997). *Les jardins ordinaires.* Paris, L'Harmattan.

El Faïz M. (2016). « L'eau fondatrice des jardins arabo-musulmans », In Coll. *Jardins d'Orient, de l'Alhambra au Taj Mahal*, Gand, Snoeck-Institut du monde arabe.

Faye A. (2017). *Analyse statistique d'une enquête sociologique sur la lutte anti-vectorielle.* Mémoire de stage de Master 1, Mathématiques appliquées et Sciences sociales, Aix-Marseille Université.

FRILEUX P. (2010). « À l'abri de la haie dans le bocage pavillonnaire », *Ethnologie française*, 40(4), 639-648.

GUSTAVE J., FOUQUE F., CASSADOU S. *et al.* (2012). "Increasing role of roof gutters as Aedes aegypti (*Diptera: Culicidae*) breeding sites in Guadeloupe (FrenchWest Indis) and consequences on dengue transmission and vector control", *Journal of Tropical Medecine*, Article ID 249524, 3 p.

HOUNKPE J. (2012). *Gîtes larvaires d'Aedes albopictus dans le bâti et les ouvrages de gestion des eaux pluviales: état des lieux et enjeux en termes de stratégie de contrôle.* Mémoire de l'EHESP, Ingénieur du Génie Sanitaire, Rennes.

JAMES S., COLLINS F.H., WELKHOFF P.A., EMERSON C., GODFRAY H.C.J., GOTTLIEB M., GREENWOOD B., LINDSAY S.W., MBOGO C.M., OKUMU F.O., QUEMADA H., SAVADOGO M., SINGH J.A., TOUNTAS K.H. and TOURÉ Y.T. (2018). "Pathway to deployment of gene drive mosquitoes as a potential biocontrol tool for elimination of malaria in Sub-Saharan Africa: Recommendations of a scientific working group", *The American Journal of Tropical Medicine and Hygiene*, 98(6), Suppl, 1-49. doi:10.4269/ajtmh.18-0083

JANSEN C.C. and BEEBE N.W. (2010). "The dengue vector Aedes aegypti: what comes next", *Microbes Infect.*, 12(4), 272-279.

LARBEY V. (2013). *Jardins et jardiniers. Les pieds dans la terre, la tête dans les nuages. Une anthropologie du potager*. Thèse d'anthropologie, Université Montpellier 3.

MARC J.V. (2011). « Le jardin créole à Fort-de-France : stratégie de résistance face à la pauvreté ? », *VertigO – la revue électronique en sciences de l'environnement*, 11, 1. En ligne : http://vertigo.revues.org/10804.

MARCO A., BARTHELEMY C., DUTOIT T. and BERTAUDIÈRE-MONTES V. (2010). "Bridging Human and Natural Sciences for a Better Understanding of Urban Floral Patterns: the Role of Planting Practices in Mediterranean Gardens", *Ecology and Society*, 15(2), 2. Online: http://www.ecologyandsociety.org/vol15/iss2/art2/

MARTINEZ ALIER J. (2014), *L'écologisme des pauvres. Une étude des conflits environnementaux dans le monde*. Paris, Les Petits Matins, Institut Veblen.

MENOZZI M.-J. (2007). « " Mauvaises herbes ", qualité de l'eau et entretien des espaces », *Natures Sciences Sociétés*, 15(2), 144-153.

MICHEL C. and THEULIÈRE M. (2010). « Les inégalités de revenus entre les DOM et la métropole », *INSEE Première*, 1279, février.

Mieulet E. and Claeys C. (2016). « (In)acceptabilités environnementales et/ou sanitaires : dilemmes autour de la démoustication du littoral méditerranéen français », *VertigO – la revue électronique en sciences de l'environnement*, 16, 1. En ligne : http://vertigo.revues.org/16940

Mieulet E. (2015). *La prolifération de moustiques vecteurs sur le littoral méditerranéen et dans les départements d'Amérique: enjeux environnementaux et sanitaires.* Thèse de Sociologie, Aix-Marseille Université.

Mounet C. (2012). « Conflits et reconfigurations socio-spatiales autour du sanglier. Des postures aux arrangements locaux », *Économie rurale*, 327-328, 79-95. En ligne : http://economierurale.revues.org/3367.

Poulin B., Lefebvre G., Muranyi-Kavacs C., Lavallée-Chouinard C. and Hilaire S. (2016). *Suivi de l'efficacité d'une démoustication par pièges Techno Bam. Hameau du Sambuc.* Rapport de recherche PNRC. En ligne : https://tourduvalat.org/.

Reiskind M.H., Zarrabi A.A. and Lounibos L.P. (2010). "Invasive leaf resources alleviate density dependence in the invasive mosquite, Aedes albopictus", *Biol Invasions*, 12, 2319-2328.

Robbins P. (2007). *Lawn People. How Grasses, Weeds, and Chemicals Make Us Who We Are.* Philadelphia: Temple University Press.

Samson D.M., Qualls W.A., Roque D., Naranjo D.P., Alimi T., Arheart K.L., ... and Xue R.D. (2013). "Resting and energy reserves of Aedes albopictus collected in common landscaping vegetation in St. Augustine, Florida", *Journal of the American Mosquito Control Association*, 29(3), 231-236.

Setbon M., Raude J. *et al.* (2008), « Les Martiniquais face à la dengue : résultats d'une étude sociologique sur les représentations, les attitudes et les comportements de la population », *Bulletin d'alerte et de surveillance Antilles Guyane*, 7, 1-10.

Souami T. (2012). « Récupération et utilisation de la pluie : un éclairage sur l'enfermement décisionnel et économique de la gestion de l'eau », *Flux*, 88(2), 7-18.

Soulancé D., Gaimard M., Bley D. et Vernazza-Licht N. (2011). « Lieux de vie et santé des populations : l'exemple du chikungunya à la Réunion », *Cahiers de géographie du Québec*, 55(156), 603-621.

Walther G.R., Roques A., Hulme P.E., Sykes M.T., Pyšek P., Kühn I., Zobel M., Bacher S., Botta-Dukát Z., Bugmann H., Czúcz B.,

DAUBER J., HICKLER T., JAROŠÍK V., KENIS M., KLOTZ S., MINCHIN D., MOORA M., NENTWIG W., OTT J., PANOV V.E., REINEKING B., ROBINET C., SEMENCHENKO V., SOLARZ W., THUILLER W., VILÀ M., VOHLAND K. and SETTELE J. (2009). "Alien species in a warmer world: risks and opportunities", *Trends in Ecology and Evolution*, 24(12), 686-693.

WINDBICHLER N., MENICHELLI M., PAPATHANOS P.A., THYME S.B., LI H., ULGE U.Y., HOVDE B.T., BAKER D., MONNAT JR R.J., BURT A. and CRISANTI A. (2011). "A synthetic homing endonuclease-based gene drive system in the human malaria mosquito", *Nature*, 473, 212-215.

Synthetic Biology and Malaria Control: Navigating between Biology and Social Science

Christophe Boëte

With the advent of molecular biology and its advances in the early nineties, researchers in the field began to consider its potential utility in the fight against vector-borne diseases. A pioneering 20-year plan was drawn up to develop genetically modified mosquitoes and conduct controlled experiments to test how an allele of interest (responsible for malaria refractoriness in the vector) could spread in wild populations of *Anopheles*. While some of the plan's milestones were reached more than 10 years ago (notably the stable transformation of the *Anopheles* mosquito and the engineering of a mosquito unable to transmit malaria), the recent development of synthetic biology has accelerated progress towards this high-tech product being added to the vector control toolbox. The discovery of CRISPR-Cas 9, often described as molecular scissors for DNA, has largely facilitated the development of gene drive systems (Gantz *et al.*, 2015; Hammond *et al.*, 2016) and has thereby made it potentially much easier to alter populations. Essentially, gene drives allow the rapid spread of a DNA cassette into a target population via sexual reproduction. A gene drive can copy and paste itself within the genome and this property gives it a more than 90% chance of being transmitted to the next generation. In theory, the release of a few individuals carrying such a construct could lead to the modification of the entire target population in under 20 generations (Burt, 2003).

There are currently two main approaches to the use of gene drives and modified mosquitoes in the fight against malaria. One is based on modifying mosquitoes to make them *Plasmodium*-refractory and focuses on the Asian malaria vector *Anopheles stephensi* (Isaacs *et al.*, 2012). The other approach aims to reduce or eliminate populations of the African malaria vector via the disruption of genes that are essential for female fertility. This project, called Target Malaria, is hosted at Imperial College London (UK). In addition to being well advanced at the theoretical level (Burt, 2003) and in the lab (Hammond *et al.*, 2016; Kyrou *et al.*, 2018), Target Malaria has recently made progress in the field with the release of modified mosquitoes (not carrying the gene drive construction) in Burkina Faso (Info'GM, 2018).

Although the technology is particularly new and disruptive, GM mosquitoes have been released in the past in several countries to reduce the population of *Aedes* mosquitoes, though without yet using gene drive. This approach was led by a British biotech company named Oxitec (Harris *et al.*, 2011). At the time, the test releases of GM *Aedes aegypti* in the Cayman Islands and later Malaysia led to controversy over the importance of public involvement and open dialogue with the communities concerned, even among researchers active in the field (Enserink, 2010). Target Malaria's more recent approach seems different. The consortium includes scientists from malarious countries and claims to engage the public in the project. It is also active in producing guidelines and recommendations for the safe, ethical testing and implementation of gene drive technology in mosquitoes to reduce malaria transmission in Africa (James *et al.*, 2018). While this apparently shows commendable concern for the social, legal and ethical aspects surrounding the potential use of "gene-drive-carrying" mosquitoes in Africa, the fact that the consortium organizes communication events in Africa in partnership with the New Partnership for Africa's Development (NEPAD), the Foundation for the National Institutes of Health (FNIH) and the International Life Sciences Institute (ILSI) is problematic. ILSI, which describes itself as non-profit organization that provides science to improve human health and safeguard the environment, is in fact a lobby group that has, among other things, been very active in promoting GM crops in the European Union (Boëte, 2018a). Moreover, it receives a large fraction of its annual budget from companies such as Bayer-Monsanto, BASF, Nestlé and Syngenta. This obviously raises questions about conflict of interest and highlights the need for a contradictory and genuinely open debate involving independent institutions.

This need appears all the more pressing following the revelations of the "gene drive files" – a large number of emails and documents released under a Freedom of Information Act (FOIA) request from Edward Hammond/ Third World Network and in response to an Access to Information request filed in Canada by ETC Group. (http://genedrivefiles.synbiowatch.org). These documents revealed efforts by the FNIH and the Bill and Melinda Gates Foundation to influence UN agencies' support of gene drive research and "fight back the gene drive moratorium proponents before the Convention on Biological Diversity CBD meeting in 2018" (Boëte, 2018b). Conducting covert activities that aim to oppose the democratic will of dozens of organizations which have called for a moratorium on gene drive research and use (http://www.synbiowatch.org/gene-drives/gene-drives-moratorium) does not help build trust between scientists and the public. It instead inspires the bitter feeling that while claiming to back the establishment of guiding principles for the sponsors and supporters of gene drive research (Emerson *et al.*, 2017), several organizations with a huge stake in the potential future use of gene drive in Africa are willing to use all means at their disposal to secure public approval.

Due largely to attitudes such as these, the gene drive debate is currently heavy with questions and suspicion. This is detrimental for research and for science in general. As Darryl Macer stated years ago, a two-way dialogue should be established between scientists and the public (Macer, 2006) in order to instil mutual confidence and make sure that innovation and new tools are co-developed. It is only by doing so that they can ultimately meet both researchers' expectations and the aspirations and needs of the communities concerned.

References

Boëte C. (2018a). "Public engagement and communication: who is in charge?", *EMBO rep*, 19, 1-2. doi:10.15252/embr.201745379

Boëte C. (2018b). "Letter: Gene Drive and Trust in Science", *Genewatch USA – Council for Responsible Genetics*, 31(1), 18-19.

Burt A. (2003). "Site-specific selfish genes as tools for the control and genetic engineering of natural populations", *Proceedings Biological Sciences*, 270, 921-928.

Emerson C. James S., Littler K. and Randazzo F. (2017). "Principles for gene drive research", *Science*, 358(6367), 1135-1136. doi:10.1126/science.aap9026.

Enserink M. (2010). "GM Mosquito Trial Alarms Opponents, Strains Ties in Gates-Funded Project", *Science*, 330(1030). doi:10.1126/science.330.6007.1030.

Gantz V.M., Jasinskiene N., Tatarenkova O., Fazekas A., Macias V.M., Bier E. and James A.A. (2015). "Highly efficient Cas9-mediated gene drive for population modification of the malaria vector mosquito *Anopheles stephensi*". Proc Natl Acad Sci USA 112: E6736-6743.

Hammond A., Galizi R., Kyrou K., Simoni A., Siniscalchi C., Katsanos D., Gribble M., Baker D., Marois E., Russell S., Burt A., Windbichler N., Crisanti A. and Nolan T. (2016). "A CRISPR-Cas9 gene drive system targeting female reproduction in the malaria mosquito vector Anopheles gambiae", *Nat Biotechnol*, 34, 78-83.

Harris A.F., Nimmo D., McKemey A.R., Kelly N., Scaife S., Donnelly C.A., Beech C., Petrie W.D. and Alphey L. (2011). "Field performance of engineered male mosquitoes", *Nat Biotechnol*, 1034-1037. doi:10.1038/nbt.2019.

Info'GM (2018). « Burkina Faso – 10 000 moustiques OGM bientôt disséminés ». 24 septembre. En ligne: https://www.infogm.org/6631-burkina-faso-10000-moustiques-ogm-bientot-dissemines.

Isaacs A.T., Jasinskiene N., Tretiakov M., Thiery I., Zettor A., Bourgouin C. and James A.A. (2012). "Transgenic Anopheles stephensi co-expressing single-chain antibodies resist Plasmodium falciparum development", *Proc. Natl. Acad. Sci. USA*, 109, E1922-E1930.

Kyrou K.A, Hammond A.M., Galizi R., Kranjc N., Burt A., Beaghton A.K., Nolan T. and Crisanti A. (2018). "A CRISPR–Cas9 gene drive targeting doublesex causes complete population suppression in caged *Anopheles gambiae* mosquitoes", *Nature Biotechnology*. doi:10.1038/nbt.4245

Macer D. (2006). "Ethics and community engagement for GM insect vector release", In C. Boëte (ed.). *Genetically Modified mosquitoes for malaria control*. Georgetown, TX, Landes Biosciences.

ÉcoPolis

La collection ÉcoPolis est dédiée à l'analyse des changements qui se produisent simultanément dans la société et dans l'environnement quand celui-ci devient une préoccupation centrale.

L'environnement a longtemps été défini comme l'extérieur de la société, comme ce monde de la nature et des écosystèmes qui sert de soubassement matériel à la vie sociale. Les politiques d'environnement avaient alors pour but de « préserver », « protéger », voire « gérer » ce qui était pensé comme une sorte d'infrastructure de nos sociétés. Après quelques décennies de politique d'environnement, la nature et l'environnement sont devenus des objets de l'action publique et il apparaît que c'est dans un même mouvement que chaque société modèle son environnement et se construit elle-même. Cette dialectique sera au centre de la collection.

Directeur de collection : Marc Mormont,
Professeur à la Fondation universitaire luxembourgeoise
(Université de Liège, Belgique)

Dans la collection

N° 31 – Cécilia Claeys (ed.), *Mosquitoes management. Between environmental and health issues*, 2019.

N° 30 – Sylvain Guyot, *La nature, l'autre frontière. Fronts écologiques au Sud (Afrique du Sud, Argentine, Chili)*, 2017.

N° 29 – Divya Leducq, Helga-Jane Scarwell et Patrizia Ingallina (dir.), *Modèles de la ville durable en Asie. Utopies, circulation des pratiques, gouvernance*, 2017.

N° 28 – Ludovic Ginelli, *Jeux de nature, natures en jeu. Des loisirs aux prises avec l'écologisation des sociétés*, 2016, 240 pages. N° 27 – Xavier Arnauld de Sartre, *Agriculture et changements globaux. Expertises globales et situations locales*, 2016, 204 pages.

N° 26 – Bernard HUBERT et Nicole MATHIEU (dir.), *Interdisciplinarités entre Natures et Sociétés. Colloque de Cerisy*, 2016, 396 pages.

N° 25 – Arnaud BUCHS, *La pénurie en eau est-elle inéluctable ? Une approche institutionnaliste de l'évolution du mode d'usage de l'eau en Espagne et au Maroc*, 2016, 331 pages.

N° 24 – Valérie DELDRÈVE, *Pour une sociologie des inégalités environnementales*, 2015, 243 pages.

N° 23 –Zhour BOUZIDI, *Se coordonner dans un périmètre irrigué public au Maroc.* Contradictio in terminis ?, 2015, 373 pages.

N° 22 – Laura SILVA-CASTAÑEDA, Étienne VERHAEGEN, Sophie CHARLIER, An ANSOMS (dir.), *Au-delà de l'accaparement. Ruptures et continuités dans l'accès aux ressources naturelles*, 2014, 244 p.

N° 21 – Xavier ARNAULD DE SARTRE, Monica CASTRO, Simon DUFOUR, Johan OSZWALD (dir.), Political ecology *des services écosystémiques*, 2014, 288 pages.

N° 20 – Céline GRANJOU, *Micropolitiques de la biodiversité. Experts et professionnels de la nature*, 2013, 202 pages.

N° 19 – Corinne LARRUE (dir.), *Le régime institutionnel d'une nouvelle ruralité. Analyses à partir des cas de la France, des Pays-Bas et de la Suisse*, 2013, 214 pages.

N° 18 – François BERTRAND et Laurence ROCHER (dir.), *Les territoires face aux changements climatiques. Une première génération d'initiatives locales*, 2013, 269 pages.

N° 17 – Véronique ANCEY, Isabelle AVELANGE, Benoît DEDIEU (dir.), *Agir en situation d'incertitude en agriculture. Regards pluridisciplinaires au Nord et au Sud*, 2013, 419 pages.

N° 16 – N° 16 – Cécilia CLAEYS and Marie JACQUÉ (eds.), *Environmental Democracy Facing Uncertainty*, 2012, 185 pages.

N° 15 – Josiane STOESSEL-RITZ, Maurice BLANC, Nicole MATHIEU (dir.), *Développement durable, communautés et sociétés. Dynamiques socio-anthropologiques*, 2012, 230 pages.

N° 14 – Philippe HAMMAN, Christine BLANC et Cécile FRANK, *La négociation dans les projets urbains de tramway. Éléments pour une sociologie de la « ville durable »*, 2011, 246 pages.

N° 13 – Denise VAN DAM, Michel STREITH et Jean NIZET (dir.), *L'agriculture bio en devenir. Le cas alsacien*, 2011, 140 pages.

N° 12 – Philippe HAMMAN et Jean-Yves CAUSER (dir.), *Ville, environnement et transactions démocratiques. Hommage au Professeur Maurice Blanc*, 2011, 291 pages.

N° 11 – Géraldine FROGER (dir.), *Tourisme durable dans les Suds ?*, 2010, 316 pages.

N° 10 – Muriel MAILLEFERT, Olivier PETIT et Sandrine ROUSSEAU (dir.), *Ressources, patrimoine, territoires et développement durable*, 2010, 283 pages.

N° 9 – Philippe HAMMAN et Christine BLANC, *Sociologie du développement durable urbain. Projets et stratégies métropolitaines françaises*, 2009, 260 pages.

N° 8 – François MÉLARD (dir.), *Écologisation. Objets et concepts intermédiaires*, 2008, 214 pages.

N° 7 – David AUBIN, *L'eau en partage. L'activation des règles dans les rivalités d'usages en Belgique et en Suisse*, 2007, 247 pages.

N° 6 – Géraldine FROGER (dir.), *La mondialisation contre le développement durable ?*, 2006, 315 pages.

N° 5 – Laurent MERMET (dir.), *Étudier des écologies futures. Un chantier ouvert pour les recherches prospectives environnementales*, 2005, 411 pages.

N° 4 – Jean-Baptiste NARCY, *Pour une gestion spatiale de l'eau. Comment sortir du tuyau ?*, 2004, 342 pages.

N° 3 – Pierre STASSART, *Produits fermiers : entre qualification et identité*, 2003, 424 pages.

N° 2 – Cécilia CLAEYS-MEKDADE, *Le lien politique à l'épreuve de l'environnement. Expériences camarguaises*, 2003, 245 pages.

N° 1 – Edwin ZACCAÏ, *Le développement durable. Dynamique et constitution d'un projet*, 2002 (2e tirage 2003), 358 pages.

www.peterlang.com

FSC
www.fsc.org
MIX
Papier | Fördert gute Waldnutzung
FSC® C083411

Kolibri 360 GmbH
Ferdinand-Jühlke-Straße 7
99095 Erfurt, Deutschland
produktsicherheit@kolibri360.de